Felix M Philip
Bosco Paul Alapatt
Anupama Jims

Real Time Multi Object Detection and Tracking Using Deep Learning

Felix M Philip
Bosco Paul Alapatt
Anupama Jims

Real Time Multi Object Detection and Tracking Using Deep Learning

Noor Publishing

Imprint
Any brand names and product names mentioned in this book are subject to trademark, brand or patent protection and are trademarks or registered trademarks of their respective holders. The use of brand names, product names, common names, trade names, product descriptions etc. even without a particular marking in this work is in no way to be construed to mean that such names may be regarded as unrestricted in respect of trademark and brand protection legislation and could thus be used by anyone.

Cover image: www.ingimage.com

Publisher:
Noor Publishing
is a trademark of
International Book Market Service Ltd., member of OmniScriptum Publishing Group
17 Meldrum Street, Beau Bassin 71504, Mauritius
Printed at: see last page
ISBN: 978-620-0-78152-9

REAL TIME MULTI OBJECT DETECTION AND TRACKING USING DEEP LEARNING

Dr. Felix M Philip is an Assistant Professor in Computer science and IT department at JAIN (Deemed-to-be University), Kochi. He has more than 9 years of teaching experience. His research interest includes Image Processing, Internet of Things.

Bosco Paul Alapatt (Ph.D.) is an Assistant Professor in Computer science and IT department at JAIN (Deemed-to-be University), Kochi. He has more than 17 years of teaching experience. His research interest includes Networking, Database.

Ms. Anupama Jims MSc in Information Science from University of Konstanz, Germany. At present, she is an Assistant Professor in Computer science and IT department at JAIN (Deemed-to-be University), Kochi. She has more than 5 years of teaching experience. Her research interest includes Data Science, Machine Learning and Educational data mining.

ACKNOWLEDGEMENT

A journey is pleasant when more than one person is involved. For the shaping up and completion of this book work many have directly or indirectly extended their helping hands. This is the pleasant moment to express our gratitude to all. With the greatest benevolence of the God, it was possible for us to accomplish this book.

I would like to thank all those who provide us with valuable feedback and inputs during the presentation of this book, and especially to JAIN (Deemed-to-be University), Kochi without their help and cooperation, this book would not have been released. The support, patience and inspiration that we always got from our family are something that we cherish above all. A special thanks to all our friends for their encouragement and support.

TABLE OF CONTENTS

LIST OF TABLES

TABLE NO	TABLE NAME	PAGE NO

LIST OF FIGURES

LIST OF PLATES

SYMBOLS AND ABBREVIATIONS

SYMBOL	DEFINITION
ψ	Mother wavelet
α	Scaling parameter
β	Shifting parameter
Δ_{ij}	Distance measure
$\mod(lo_i)$	The mode of the location lo_i
Th_a	Threshold area
A_i	Area of the object pixel
o_i	Object pixel
$E_L(i)$	Left edge of the object image o_i.
$E_T(i)$	Top edge of the object image o_i
$E_B(i)$	Bottom edge of the object image o_i.
D	PPCM density
$\{O_b^1, O_b^2, O_b^3, \ldots\ldots, O_b^m\}$	Array of objects
$X_R(i)$	Initial position of image
$U_R(i)$	Extended edges of image
V	Input video

ACRONYMS   **ABBREVIATIONS**

DVD	Digital Versatile Disk
HDTV	High Definition Television
DSS	Digital Satellite Systems
SMC	Sequential Monte Carlo
DBT	Detection Based Tracking
DFT	Detection Free Tracking
MCMC	Markov chain Monte Carlo
MTVT	Multi-target Visual Tracking
MOT	Multiple Object Tracking
HCI	Human Computer Interface
VAR	Virtual Augment Reality
HINE	Hammersmith Infant Neurological Examination
SCP	Shape Control Points
MDA	Multi-Dimensional Assignment
PPI	Pixels Per Inch
PPCM	Pixels Per Centimeter
TWSM	Tangential Weighted Spatial Model
SDVM	Second Derivative Visual Method
GM-PHD	Gaussian Mixture Probability Hypothesis Density
MOTSFK	Multiple Object Tracking By Employing Shaped Based Features And Kalman Filter

CHAPTER 1

INTRODUCTION

Video processing is a technique used for examining video contents in order to get an idea of the scene that the video represents. Video analytic methods are inspired by the requirement of generating machine algorithms which imitate the abilities of humans and other living organism visual systems. Video analysis is an important part of various technologies such as robotics, video surveillance, and multimedia. There are a number of research works for video analysis carried out in various fields like computer science, statistics, signal and image processing, and system theory. The Video processing method has created changes in multimedia applications with products including Digital Versatile Disk (DVD), High Definition Television (HDTV), video cameras, and Digital Satellite Systems (DSS). There are four fields in video processing. They are video compression, video indexing, video segmentation, and video tracking.

1.1 TRACKING

Due to the improvement in sensor models, tracking systems play an important role in video analysis. Tracking methods detect the objects in the video and link the results with the object trajectories. The tracking system is used to initialize and terminate the trajectories by using the appearance and discriminative information, and attains state-of-the-art performance (Xiao Liu *et al.*, 2014). Tracking methods consists of two steps: the first one is the global data association and the second one is the Sequential Monte Carlo (SMC) process. In the global data association algorithm, visual proofs are collected from the past and future frames over a broad time window and the results are combined with track lets. The global data association algorithm did not perform online processing. On the other hand, the Sequential Monte Carlo process uses only the visual proofs from the past frames and it is useful for real-time applications. The SMC process continuously generates multiple hypotheses from the past states to get the exact associations with noise detection responses (Xiao Liu *et al.*, 2014).

1.2 OBJECT TRACKING

Object tracking plays an important role in vision systems. The main aim of object tracking is to find the location of the object in every frame of the video (Alper *et al.*, 2006). The applications of object tracking include traffic pattern analysis, video surveillance, robotics, and human-computer interaction. The two methods in a visual tracker are: 1) Target representation and localization, and 2) Filtering and data association. The changes in the target appearance are handled by the target representation and localization method and the learning of previous scenes, and hypotheses evaluation are handled by the filtering and data association method (Comaniciu *et al.*, 2003 and Dae-Hwan *et al.*, 2014).

1.2.1 Object Representation

Every object is represented by its shape and appearance. Object representation based on shape is described in the following:

Points: Here, the objects are represented by a point which means centroid (Veenman *et al.,* 2001) or a group of points (Serby *et al.*, 2004). Point representation is used for object tracking in an image where the object occupies small regions.

Primitive geometric shapes: Here, the object is represented by a primitive geometric shapes such as an ellipse (Comaniciu *et al.*, 2003), rectangle and so on. Affine, translation, and projective (homographic) transformation are used to model the motion of the object in this type of representation. Primitive geometric shapes are useful for both rigid and non-rigid objects.

Object silhouette and contour: Contour representation is used to define an object's boundary. The object silhouette represents the region which is inside the contour. The object silhouette and contour representation are used for tracking non-rigid objects (Yilmaz *et al.*, 2004).

Articulated shape models: Articulated objects are objects, in which parts of the body are linked together by joints. An example of an articulated object is the human body where the different parts such as legs, hands, torso, feet, and head are connected by joints. The parts of the body are related by kinematic motion models such as a joint angle. An articulated object is represented by cylinders or ellipses.

Skeletal models: The medial axis transform is applied to the object silhouette (Ballard *et al.*, 1982) from which the object skeletal is extracted. Skeletal models are used to represent the shape of the objects for identifying them (Ali *et al.*, 2001). Skeletal models are worthy for representing rigid and articulated objects (Alper *et al.,* 2006). Appearance based object representations are described in the following sections:

Probability densities of object appearance: The evaluation of the probability densities of the object appearance is either parametric or non-parametric. Here, features such as colour and texture are calculated from the regions of the image defined by the shape models.

Templates: Templates are used to encode the appearance of the object produced by a single view and it is only worthy for tracking objects when there is no change in the position of the object during the time of tracking. Object silhouettes or primitive geometric shapes are used to define the templates (Fieguth *et al.,* 1997) Templates consider both the appearance and spatial information of the objects, which is a major benefit of the template.

Active appearance models: The shape and appearance of the objects are modelled to generate the active appearance models (Edwards *et al.,* 1998). The shape of the objects is described by a group of landmarks. The landmark resides on either the boundary of the objects or inside the object region. The colour, gradient magnitude or texture is stored as the appearance vector for every landmark. In the training phase of the active appearance model, the shape and appearance of the object is cultivated from the set of samples.

Multi-view appearance models: In Multi-view appearance models, the subspaces are generated from the given view of the object to represent the object in different views. Subspace methods are used to represent the shape and appearance of the object. Examples of subspace methods are the Principal Component Analysis (PCA) and Independent Component Analysis (ICA) (Mughadam *et al.,* 1997, Black *et al.,* 1998, Alper *et al.,* 2006)

1.2.2 Object Detection

The object detection mechanism plays an important role in each and every tracking method, in which the detection is performed in every frame of the video or the frame where the object become visible at first. Generally, the object detection is performed based on the information presented in a single frame of the video. Some of the object detection methods use the temporal information to minimize false detection. The temporal information obtained from the series of frames has been used to highlight the object regions which are changing in successive frames and the temporal information is represented as the frame differencing. The tracker identifies the similarity of the object from frame to frame in order to initiate the tracks (Alper *et al.,* 2006).

1.2.3 Types of Tracking

There are two approaches in tracking. The first one is recognition-based tracking and the second one is motion-based tracking.

1.2.3.1 Recognition-based tracking: In recognition-based tracking, the recognition of the object is modified. The object recognition is performed in consecutive images and the position of the image is identified. The major advantages of recognition-based tracking are, it evaluates the rotation and translation of the object and the tracking is performed in three dimensions. The disadvantage of this type of tracking is that it only tracks the objects that are recognizable, because the object recognition needs a high high-level operation. Therefore, recognition-based tracking systems have limited performance.

1.2.3.2 Motion-based tracking: The Motion-based tracking system detects the moving object by finding the motion of the object. The advantage of motion-based tracking is to track objects without considering the size or shape of the object. The two methods in motion-based tracking are optical flow tracking and motion-energy tracking.

1.2.3.3 Optic Flow Tracking: Optic flow tracking detects the motion of the objects from videos taken by a fully active camera. Optic flow tracking can efficiently detect the object's motion in an unconstrained domain. The limitation of this tracking is to qualitatively evaluate the motion of the object and it is more unfair than the quantitative method. The Determination of the entire optic-flow field requires high cost. Hence, the alternative approach which identifies some features of the objects in the video and tracks the motions of that objects is MOTSFK (Fieguth *et al.,* 1997, Bhanu *et al.*, 1990).

The limitation of this approach is that the features of the object in each frame must be identical to the features of the objects in the previous frame. This problem will be intensified in the videos taken by the active camera. Since the image sequence observed is dynamic, some features are moved far away from the field of vision and the new features will be moved into the field of vision which increases the strain for finding the matching features. Velocity field withdrawal is another problem in optic flow tracking and hence, optic flow tracking is not suitable for real time applications.

1.2.3.4 Motion Energy Tracking: The alternative approach for tracking the motion of an object is motion energy tracking. In this approach, the noise in the image sequence is filtered and the image is segmented into movement regions and inactivity regions, by finding the temporal derivative of an image and thresholding at an appropriate level. Generally, the temporal derivative can be calculated by straightforward image subtraction, which creates noise and produces vague values. The performance of the image subtraction can be improved by using spatial edge information for finding the moving edges. The most successful and widely used method for motion detection combines image subtraction and spatial information. Motion energy tracking is computationally very simple and it is worthy for pipeline architectures. The limitation of motion energy tracking is that it detects the motion of the pixel but does not measure the quantity. Hence, additional information like the focus of expansion cannot be determined. This type of tracking is not applicable for videos from an active camera. The active camera systems persuade the obvious motion of the objects in the image sequence, so that the indemnity for the obvious motion is made prior to the motion energy tracking (Murray *et al.,* 1994).

The general object tracking techniques use rough shapes like ellipses and rectangles for representing the objects (Zhou *et al.,* 2007), while the active contour-based tracking (Cremers *et al.,* 2007) presents the more comprehensive representation of the object. Active contour-based tracking is harder than the general object tracking techniques for tracking the same object, because contour-based tracking tries to find the detailed representation of the objects like object boundary and object boundary detection which is harmed by the effects from the background disruption. In videos captured by stationary cameras, the object movement regions are determined by the background subtraction process and the edges of the movement regions are discovered to produce the object contours. On the other hand, in videos captured by the non-stationary cameras, background subtraction is not used for finding the object movement regions, which makes the active contour-based tracking harder for videos captured by the non-stationary cameras. However, active contour-based tracking is a more powerful approach for object tracking in recent years (Weiming *et al.,* 2013).

Object contours are represented by two types: The first is the explicit representation and the second is the implicit representation. The explicit representation describes the features of the parameterized curves like snakes (Kass *et al.,* 1988) and the implicit representation describes the contour by a signed distance map like level sets (Osher *et al.,* 1988). Compared to the explicit representation, the implicit representation is favourable because it produces a strong numerical solution and it performs well even if there are topological changes.. Active contour-based tracking is divided into three types of methods. They are, 1) edge-based, 2) region-based, and 3) shape prior-based methods.

1) Edge-Based Methods: The Edge-based method examines the local information like the grey level gradient throughout contours for representing the object contours. An example for the edge-based method is the snake model. The edge-based methods are simple and effective for identifying the object contours. The drawbacks of the edge-based methods are described in the following: The first contour must be described close to the objects and the object contours described in the analogous regions need not be optimized.

2) Region-Based Methods: Depending on the principle of statistic quantities like variance and mean, the region-based methods separate the objects and the framework from the image. In region-based methods, the past knowledge about the object's colour and texture are combined for evaluating the contour. The object appearance models are used to describe the previous knowledge about the object colour. Examples of object appearance models are colour histograms, and Gaussian Mixture Models (GMMs). The major advantages of region-based methods are robustness and correctness. The disadvantage of the region-based method is that it assumes that the pixel values are independent of each other for estimating the posterior probability function (Cremers *et al.,*2007) so that the evaluated contour is easily affected by the disruptions of texture or colour similarities across the background and object.

3) Shape Prior-Based Methods: Shape prior-based methods are used to model the shape of the objects according to statistics and recover the object contours which are affected by occlusion, and blurring. Adaptive shape-based methods (Paragios *et al.,* 2002) twist the contour sections that are undisturbed and determine them by the colour features; on the other hand, they determine the disturbed contour sections globally. The active shape prior-based methods need continuous updating to adopt, according to the changes in the object shape. The recent method (Fussenegger *et al.,* 2006), which updates the shape prior-based methods does not process the multiple modern shapes. A simple data fitting process for regular movements of non-rigid objects described in (Cremers 2006) is the dynamic shape model which has no knowledge about changes in shape. This model imagines that the hidden movement of the object is close to the periodic motion of the object (Weiming *et al.,* 2013).

There are two types of approaches in visual tracking. The first one is the bottom-up approach and the second one is the top-down approach. Bottom-up approaches are used to restore the target by evaluating the contents of the image. An Example of bottom-up tracking is the reconstruction of a parametric shape through the use of curve fitting. On the other hand, top-down approaches produce and assess state hypotheses depending on the target states and track the target by assessing and checking these hypotheses on observations of the image. The robustness of the bottom-up approaches depends on the image analysis, because activities such as grouping, and tracing are deluged by the noise and clutters in the image, and they are very efficient. In contrast, the top-down approaches are slightly based on image analysis because the image can be analysed based on the target hypotheses and the performance is calculated by hypotheses generation and verification. The top-down approach requires a huge number of hypotheses in order to attain robustness. Hence, it requires additional computation for hypotheses evaluation. In some situations, the top-down and bottom-up approaches are mingled to increase the robustness. At the same time, the computation can be reduced (Ying *et al.,* 2004).

An adaptive tracking method improves the tracker's performance by adaptively choosing the features of the object that differentiate the object from its background. One way to increase the performance of the tracker is to combine the different cues such as texture, shape and colour (Junqiu *et al.,* 2008).

1.2.4 Elements of Visual Tracking

The four elements in visual tracking are: 1) target representation, 2) observation representation, 3) hypotheses generation, and 4) hypotheses evaluation.

1.2.4.1 Target representation: Target representation is used to differentiate the target from the other objects implicitly or explicitly, by considering the shape, appearance, colour, and motion of the objects and it is denoted by X. In computer vision and visual tracking, target representation is the major problem, but it provides a succinct representation of the target object and achieves high computational efficiency. The unique representation of the target will ease the visual tracking methods, and at the same time include high dimensionality. Several methods use the appearance of the image like image templates (Hager *et al.,* 1996, Li *et al.,* 2000, Tao *et al.,* 2000), and Eigen-space representation (Zhou *et al.,* 2007) for providing a unique representation of the target. Example, the person in a crowd is tracked easily if he/she is already known.

Observation representation: Observation representation is used to define the evidence of the image which means the features of the image and it is denoted by Z. When the representation of the target is defined by the contour shape, the edges of the corresponding image need to be defined in the observation representation and when the target is represented by the colour appearance, the colour distribution patterns of the corresponding image need to be defined in the observation representation.

1.2.4.2 Hypotheses evaluation: The hypotheses evaluation is used to define the matches across the image observation and state hypotheses. In hypotheses evaluation, the likelihood region of the image observations is measured. If the image contains clutter, the hypotheses evaluation process becomes a challenging task for accessing a state hypothesis.

1.2.4.3 Hypothesis generation: Hypotheses generation is used to create new state hypotheses by evaluating the previous estimates of the target state and it is denoted by p $(X_t | X_{t-1})$, where X_t and X_{t-1} are the present and previous target states respectively. Hypotheses generation is also used to intimate how the dynamic process of the target is developed and the dynamic process of the target is embedded in a predicting process. At a particular time instant, the target state of the tracking process is a random vector and it can be computed by the conditional probability density function. Hence, the range of search and level of confidence of the tracking process are characterized by the hypothesis generation process. The robustness of the tracking process is improved by recruiting more target representation, because the target representation process is used to provide the uniqueness of the target state (Ying *et al.*, 2004).

1.3 MULTI-OBJECT TRACKING

One of the major problems in video analysis is multi-object tracking. Multi-object tracking is the process of finding multiple objects from the input video in the 2D image plane or 3D object space. Generally, the tracking process is represented as a first-order Markov chain process, in which the location of the objects at a certain time step is calculated by those objects in the past time step. Then, the object models are compared to the current image data to refine the tracking process and update the object models. The same procedure is performed at the next time step. From the Markov chain process, it can be understood that once the tracker loses the track, the wrong information is obtained for the next time step. Therefore, the tracker cannot regain from failure. This is considered as a major problem in multi-object tracking in which object occlusions and object-object interactions are common (Leibe *et al.,* 2008). At first, multi-object tracking detects the objects in the individual frames of the input video and links the detected objects between the frames. The two steps in multi-object tracking are time-independent detection and modelling detection. Time-independent detection deduces the location and number of targets from the input video at every time step. This step requires a generative model or a discriminative machine learning based algorithm. Modelling detection is used to find the errors and target motions, in order to connect the detections with the most plausible trajectories (Jerome *et al.,* 2011).

Multi-object tracking is performed on videos caught by stationary or non-stationary cameras.

1.3.1 Multi-Object Tracking with Stationary Cameras

For tracking multiple objects from the videos caught by the stationary cameras, the past information regarding the locations of moving objects is obtained based on the image calibration, homographic constraints across the multi-cameras, background subtraction and so on (Wang *et al.,* 2005, Song *et al.,* 2008).

1.3.2 Multi-Object Tracking with Non-stationary Cameras

For tracking multiple objects from videos taken by the non-stationary cameras, calibration, homographic constraints across the multi-cameras, and background subtraction are not required for finding the past information of the locations of the moving objects. Hence, multi-object tracking in videos taken by non-stationary cameras is harder than the multi-object tracking in videos taken by stationary cameras. The major disadvantage of multi-object tracking with non-stationary cameras is that it assumes, that the appearance of the objects is not changed during occlusion. In the existence of object occlusions, the appearance of the object is changed more and the tracking process does not track the objects correctly (Weiming *et al.,*2012)

1.3.3 Terminologies in Multi-object Tracking

The terminologies used in multi-object tracking are object, detection, tracking, detection response, trajectory, tracklet, and data association.

Object: An object can be anything in an image, represented as a closed area and the object can be distinguished from the background. The object tracking process detects the identity of the objects along with the object and the absorption of multi-object tracking is inspired by tracking objects of the same type.

Detection: Detection is the process of identifying the objects from the image. The object detector is used to detect the object from the image which was in the training data set. The object detector is instructed for a particular kind of objects such as vehicle, human, and so on. The object detection process did not identify the temporal information of the objects in most of the circumstances.

Tracking: Tracking is the process of localizing a similar object in consecutive frames of the video. Object tracking requires the temporal information of the object. Multi-object tracking is used to identify the multiple objects from the image sequence and manage the identities of the objects.

Detection response: The other names of detection response are detection observation and detection hypotheses. Detection response is the results obtained from the object detector representing the appearance of the objects such as size and position of the objects in the video.

Trajectory: The output of the multi-object tracking system is the trajectory. Each target has a unique trajectory which represents the appearance of the object in one frame. The trajectory is created by combining the object responses of similar objects in the video.

Tracklet: Tracklet is the results obtained in-between the detection response and trajectories, and it is created by combining more detection responses from a similar target of an image sequence. A detection response is the tracklet with one detection response. A Tracklet is smaller than a trajectory in respect of time span because it is acquired by connecting the self-assured detection responses.

Data association: Data association is the process, in which the detection responses are matched between frames depending on the object detection. Data association is used to represent the inter-frame similarity across the detection hypothesis.

1.3.4 Multi-Object Tracking Categorization
1.3.4.1 Initialization Method

Depending on the initialization of objects, multi-object tracking is divided into two groups. The first one is Detection Based Tracking (DBT) and the second one is Detection Free Tracking (DFT). Detection Based Tracking depends on object detection; on the other hand, Detection Free Tracking is not based on object detection.

Detection Based Tracking: At first, the detection based tracking method determines the objects in each frame and associates the object hypotheses with the trajectories. The advantage of detection based tracking is that it finds the new objects and terminates the vanishing objects automatically. Detection based tracking fails due to the following reasons: DBT methods do not focus on object detection and are constructed based on the pre-trained object detector. For the reason that DBT methods use the pre-trained object detector, they paid more attention only to particular types of targets like vehicles, pedestrians and so on. The performance of detection based tracking is based on the performance of the pre-trained object detector.

Detection Free Tracking: In detection free tracking, the finite numbers of objects are initialized in the first frame manually and these objects are identified in the successive frames of the video. The DFT method is not based on the object detector. The performance of the DFT is reduced as the conventional visual tracking problem when the video consists of only one object for tracking. In contrast to DBT, the detection free tracking method does not use the pre-trained object detector and each object to be tracked is identified manually in the first frame of the video. Therefore, DFT is suitable for tracking a sequence of objects of any type. The disadvantage of this method is that the performance of this method is limited in non-theoretical systems.

1.3.4.2 Processing Mode

Depending on the processing mode multi-object tracking is divided into two sets. The first one is online tracking and the second one is offline tracking. The difference between online tracking and offline tracking lies in the context of observation utilization. Online tracking uses the observations up to the current frame to estimate the future frame; on the other hand, offline tracking uses the observation in the past and future frames.

Online tracking: On-line tracking is also called as sequential tracking because it handles the video in a step-wise manner. Generally, online tracking is suitable for the sequential video frame.

Offline tracking: Offline tracking is also named as batch tracking because it handles the image sequence in a batch manner. Offline tracking collects the observations from all the frames of the video and analyzes these observations to evaluate the final output. The limitation in offline tracking is that it does not have the ability to consider all the frames at a time. This problem can be avoided by dividing the entire video into a number of segments and processing these segments to obtain the final results. Offline tracking acquires the global optimal solution theoretically.

1.3.4.3 Mathematical Modelling

Depending on the mathematical modelling, multi-object tracking is divided into two sets. The first one is probabilistic tracking and the second one is deterministic tracking.

Probabilistic Tracking: The probabilistic tracking method evaluates the object's states depending on the probabilistic inference and the output of probabilistic tracking varies in different running trails.

Deterministic Tracking: deterministic tracking evaluates the object's states depending on the deterministic optimization and the result of the probabilistic tracking method is constant (Wenhan *et al.,* 2015)

Multi-object tracking approaches are defined in the following three categories:

Category 1: The object tracking process and the radar signal are homogeneous. In object tracking, one-to-one mapping does not hold always, because the objects may be subjected to activities like split or merge, reappear or disappear, etc. Hence, multi-to-multi mapping is often met in the tracking process. When the number of objects increases, the hypothesis will also increase. The Growth of the hypothesis can be pruned by the use of Murty (Zhen *et al.,* 2006) and (Viterbi *et al.,* 2008).

Category 2: The particle filtering method is used for evaluating the Bayesian inference. The other name of particle filtering is the Sequential Monte Carlo method which is used for multiple object detection and tracking (Kreucher *et al.,* 2005). The boosted particle filter (BPF) (Okuma *et al.,* 2004) attains a superior performance by using the classifier.

Category 3: Another method for evaluating the Bayesian inference is the Markov chain Monte Carlo (MCMC) which is a simulation method. MCMC is developed for multiple object detection and it is used to optimize the arbitrary energy function. The methods based on MCMC are suitable for solving computer vision problems. In multi-object tracking, colour and appearance cues are used, and in radar, motion cues are used. The MCMC-particle filter (Khan *et al.* 2005a, Khan *et al.* 2005b) tracks multiple objects by using a Markov random field motion. The advantage of the MCMC based method is that it integrates the top-down and bottom-up cues. The disadvantage of the MCMC based method is that it needs a huge number of iterations for evaluating (Hai-Xia *et al.,* 2011).

1.4 MOTIVATION

Object tracking is an important problem in video analysis. One of the major problems in object tracking is the different appearance of the object and its surroundings. The target appearance will change during the prolonged tracking process, because of differences in appearance or viewpoints of the objects. Although, the video is taken by stationary cameras the background of the image will change. Object occlusion is another important challenge in the field of object tracking, in which only the regions of the occluded objects can be seen and the similarities among the objects and their features begin to be equivocal (Frey *et al.,* 2003). Occlusion reasoning is the process of discovering the occlusion relations among the objects in a detailed manner and assign the objects correctly (Weiming *et al.,* 2009).

The tracking method assumes the reliable detection of an object, but it cannot hold on every occasion. In some situations, the detector must consider the missing rate (MR) and the false positive (FP). This causes failure in tracking. Multi-target Visual Tracking (MTVT) is a critical issue in activity analysis and intelligent video surveillance systems and so on (Kosmopoulos *et al.,* 2013). An important problem in MTVT is that it needs to find the multiple number of moving targets, which are unknown and differ in time. Nowadays, a number of research works travel through the Gaussian mixture probability hypothesis density (GM-PHD) filter (Wang *et al;* 2007) for tracking multiple objects from an image sequence. During the process of data association, some difficulties may occur which can be solved by the GM-PHD filter. Still the development of the GM-PHD filter is a challenging task because of noisy data, object occlusion, and difference in the number of targets (Xiaolong *et al.*, 2014)

Multiple Object Tracking (MOT) is a difficult task in the field of visual surveillance applications, which generally come with radar applications. MOT is used to approximate the number and state of the objects or object groups from the video supplied by the sensors. Anyhow, multi-object tracking has some problems like occlusion of objects, different number of objects, object confusion, and clutter in the environment, which produces the multi to multi similarity among the tracked objects and the recently identified blobs. Single-object tracking suffers from the problem of having appearance changes, noise changes and dynamic changes in the image sequence (Hai-Xia *et al.*, 2011). The important problems in multi-object tracking are the interconnection of object detection to the object trajectories and mutual occlusion of the objects.

The drawbacks of the contour-based algorithms are described as follows: The tracking process starts with the creation of the closed contour in the surroundings of the object, and was performed manually. In this method, the motion region's boundaries are the first contours of the objects as in Paragios *et al.*, (2000) and determined by the process of background subtraction. The contour-based algorithms for object tracking are suitable only for videos captured by stationary cameras. Fast and automatic initialization of object contour in the contour-based algorithms is still a challenging problem. The level set-based methods for tracking fail to detect the object contour when there is a sudden movement of the objects. Anyhow, the sudden movements of an object are a serious problem in the contour-based tracking methods (Weiming *et al.,* 2013)

Real-time object tracking has been widely used in several applications, including human computer interface, visual surveillance applications, intelligent transportation systems, and robotics. Although, a number of methods and research works use MOTSFK for object detection and tracking from a video, still it is a difficult problem in visual applications. Object occlusion, unexpected movement of the object, and changes in appearance are the important problems in object tracking.

The advantages of multi-object visual tracking are described in the following:

Visual Surveillance: The anomalous behaviour of the objects in surveillance videos can be identified by the automatic analysis, which includes the analysis of the actions of the object and object trajectories, and so on. The goal of multi-object tracking in a surveillance video is to identify the targets from the video and track those targets.

Human Computer Interface (HCI): Human Computer Interface can be achieved by utilizing visual information like expression and movement. Visual tracking is used to extract such information. If a number of objects are visible in the video, the tracking method needs to consider the interactions between those objects. In such situations, multi-object tracking is used to achieve the human computer interface.

Virtual Augment Reality (VAR): The Virtual augment reality problem can be solved by the multiple objects tracking method, which helps users who are well experienced in video conference applications in solving this problem.

Medical Image Processing: Medical image processing needs manual labelling for performing some jobs like multiple cells labelling in images. Multi-object tracking is used to decrease the cost required for labelling (Wenhan *et al.,* 2015).

1.5 ORGANISATION OF THE THESIS

The thesis organization is done as follows,

Chapter 2 : Provides a brief survey of various existing models in object tracking in the video. The research papers with the various algorithms discussing single object tracking and multi-object tracking are summarized in this chapter.

Chapter 3 : Gives the objective of the research work.

Chapter 4 : Gives the problem statement of the work done in the research.

Chapter 5 : Introduces the MOTSFK wavelet-based object tracking algorithm for performing single object tracking in the videos. The wavelet-based object tracking performs the tracking through shot segmentation and the shape based features. It also presents the MOTSFK Kalman filter based multi-object tracking algorithm. The various feature extraction and segmentation techniques used by the MOTSFK model are discussed. The MOTSFK hybrid tracking model based on the second derivative based visibility model and the tangential weighted spatial tracking model are also mentioned in the chapter.

Chapter 6 : Concludes the research work with the summary, major findings, and the scope of future work.

CHAPTER 2

REVIEW OF LITERATURE

In this chapter, various object tracking methods have been discussed. This chapter analyses various tracking algorithms to detect the single objects and the multiple objects from the videos. This analyses helps the researcher to understand the video tracking algorithm.

2.1 VIDEO OBJECT TRACKING

The video object tracking can be broadly classified into two types. They are,

- Single object tracking
- Multi-object tracking

2.1.1 Single Object Tracking Algorithm

In the single object tracking, the tracking algorithm tracks only single target object from the video. This applies in the application where the crowd density is very low. Some of the existing methods use the single object tracking. The single object tracking uses the following algorithms,

- Filtering based techniques
- Probabilistic based techniques
- Feature components based technique
- Searching based technique

2.1.1.1 Filtering based techniques

Shao-Yi *et al.* (2013) have MOTSFK a framework for segmentation and tracking of objects in a video captured by the cameras in visual surveillance applications. In this method, the object was segmented by the threshold decision algorithm which relies on multi background registration technique (MBR eg). The threshold decision algorithm determines the thresholds vigorously for performing segmentation in the absence of user input. Therefore, this algorithm facilitates best performance for dynamic background conditions without the need for threshold tuning which was performed by the developers. The mechanism followed by this algorithm is non-identical to the background subtraction mechanism which stops the errors. Then, the object tracking was performed by the particle filtering method which was designed with diffusion distance (DD) to calculate the similarity between colour histogram and movement clues from object segmentation. Colour histogram model was included with this method to track non-rigid objects. The cross-bin distance measurements were used to address the illumination changes and diffusion distance was used to measure cross-bin distance which was worthy for hardware realization and parallel implementation. The computational complexity was reduced by using the 1-D colour histogram. This method avoids the background clutter problem by knowing the motion clues from the segmentation of objects and did not need any other computation for identifying the motion information. The drawback of this method is that it did not address issues, such as foreground clutter and object occlusions which are the common problems in object tracking mechanism.

Maha *et al.* (2014) have presented an online object tracking mechanism in particle filtering framework to track the moving objects from the video captured by the stationary cameras. The object detection mechanism uses the 'Choquet' fuzzy integral to combines the features, such as the red green blue (RGB) colour feature, the local bi nary pattern texture feature, and the Sobel edge feature to increase the correctness of the detected moving objects. It was done by took out the colour and edge grey scale confidence maps and initiate the texture confidence map. Both the continual confidence detectors and the three classifier confidence maps were used in the tracking process which are extracted from the confidence maps and an online boosting classifier respectively. At last, the Choquet integral combines the confidence detectors and the classifier maps into the particle filter framework. This method was used to track the foreground target in a more efficient manner and increase the performance of the object detection and tracking processes. At the same time, this method decreases not only the computational complexity but also the time needed for perform computation. The experimentation of this method was conducted in not only the indoor dataset sequences but also in the outdoor data set sequences. From the results of experimentation, it was found that this method tracks the moving objects from the video in an efficient manner even if variation in appearance of the objects and the movement of scene changes in an unexpected manner. Although, this method outperforms the other state-of-the-art mechanisms for object detection and tracking it had limitation that it did not works on colour spaces other than RGB.

Junda *et al* (2010) have presented an object tracking mechanism for tracking moving objects. The applications find in traffic monitoring, security surveillance and so on where the objects motion was driven by the structured environments. Hence, the relationship among the objects and its background provide the information for increasing the tracking performance and it was modelled by the environment state. The environment state was integrated with the Bayesian tracking framework and it was modelled by the distance transform. The particle filtering framework was used to solve the problem of Bayesian tracking and the particle filter was composed with environment prior and the adaptive dynamics model for the utilization of environment information. The environmental model of the structured environments was acquired by the methods of segmentation. In this method, the segmentation was performed by the mean shift algorithm in which the boundaries were refined manually to produce shiny contours for the process of distance map creation. The results of experimentation show that this method produces the better results while compared to the generic particle filtering method. The segmentation process through the manual intervention was not a matter because the distance map was used for the enlarged time period in visual surveillance. The major advantage of this method is that it tracks the movements of objects in a strong and effective manner in structured environments. The disadvantages of this method are that it did not handle object occlusion and association of data for multi-object tracking.

The background subtraction algorithm for object tracking was MOTSFK by Satrughan *et al.* (2016), which tracks object in both static and dynamic background conditions. This method founds the region of the target object with correctness and high similarity. In this method, foreground moving blobs were found by the initialization and background module updating to increase the correctness of tracking. At first, the movement field was generated on the successive frames of the video by the spatial-temporal filtering method. The moving pixels of the first movement field were extracted by evaluating the block-wise entropy on the particular pixel range of the difference image. Then, the threshold value was approximated to allocate the label to the foreground moving blobs. At last, the foreground object was determined by integrating the adaptive Kalman filter to the object extraction module and the background clutter was decreased by integrating the spatio-temporal processing and regional level operation. This method eliminates the aperture distortion in an effective way. The position of the target frames in the consecutive frames of the video was determined by providing the object's sample size to the Kalman filter. The experimentation results shows that this method track the target objects in an effective manner and increase the tracking process on both static and dynamic conditions. The result of simulation was proved that this method was implemented to track objects from a video taken by the moving or stationary cameras.

The object tracking method MOTSFK by Wu-Chih *et al.* (2015) was used for tracking objects from a video taken by the moving camera with no use of extra sensors. It was very difficult to track the moving objects from the video taken by the moving camera because the motion of camera and object were mixed. At first, this method founds the features of frame and then decides whether the features are background or foreground features. The foreground features of the frames were remunerated to acquire the updated foreground features and the ego-motion of a moving camera was remunerated to acquire the frames which are transformed through a perspective transform. Then, the image difference scheme was used for acquiring the foreground regions of the object. Then, the foreground features and the foreground regions of the objects were used to find the regions of the moving objects. The moving object regions were increased by the compensation scheme which was based on the history of movement acquired from the three successive frames of the video. Then, the minimum bounding box and the refinement scheme was used to detect the moving objects from the video sequence. At last, the moving object was tracked by the Kalman filter which depends on the gravity centre of the region of the moving object in the minimum bounding box. The results of experiments shows that this method tracks the moving objects from the video taken by the moving camera in an effective manner and track the objects which are overlapped and moves towards or away from the camera. This method was also used to track the objects which are moved fast and detect the objects from the video taken by the fast moving camera.

Xiaokai *et al.* (2016) have suggested an object tracking algorithm in a video sequences which combines the adaptive least squares and Kalman filter. The major problem in the video surveillance is the object occlusion which occurs due to problems in environment and angle. The object occlusion creates many problems to the process of video object extraction so that the tracking methods need to meet several constraints. One of the major problems in video object tracking is the selection of accurate position of the objects. In this algorithm, the frame difference method was used to find the target object and the Kalman filter was used for producing the smooth trajectory of the object. When object occlusion was encountered, this method uses the least square method to find the trajectory of the target. Hence, the exact position of the target object was determined. The experimental results shows that this algorithm can track the location of the target object even if the target is suffered from occlusion and yields correctness. This algorithm provides best tracking although the object detection results were not perfect and had maximum practical value.

Feng (2016) was presented a particle swarm optimization genetic particle filter algorithm (PSOGA) algorithm depends on local multi-zone to avoid the problems in the particle filtering framework. The particle filtering algorithm had several problems, such as particle degradation, and absence of result. At first, this algorithm moves the particles which are sampled to the high likelihood area by the particle swarm optimization algorithm and avoids the degradation of weight of the particle. Then, this algorithm makes use of the genetic algorithm to maximize the diversity of the particles and increase the algorithm's ability to perform global search. Hence, this algorithm was used to reduce the scarcity problem of the particles. This algorithm integrates the recent measurement information with the importance density function. The importance density function was generated near to the object's posteriori probability distribution. If the target object was concealed, this algorithm chooses the particle state model randomly from the local area. Therefore, the state model of particles consists of some amount of block information and avoids the barrier interfering problems. The calculation amount was reduced due to the description of objects by the local area. The experimentation of this algorithm was conducted in video sequences and the results of experiments shows that this object tracking algorithm tracks the objects in a well effective manner.

Denis *et al.* (2015) have suggested the video object tracking algorithm which was performed based on the video feed. The object detection was performed in unsupervised manner by the velocity criteria. This algorithm tracks the objects from the video captured by the single camera and Inertial Measurement Unit (IMU) sensors. In this algorithm, the object detection and tracking was augmented through the estimating the geographical coordinates. The mapping of geographical co-ordinates was used to match the position of the object on the image and the position of object on the ground. The major advantage of this algorithm was, it detects and track objects from the video taken by all the sensors and it did not limited to particular applications. Then, the tracking was performed by integrating the Bayesian filter with approximate inference. In real world co-ordinates, the localization of objects was depends on the results of tracking and IMU sensor measurements. Another advantage of this algorithm is that it provides correct tracking results. The disadvantages of this algorithm are, it requires improvements to geo-positioning, and the Bayesian filter method requires stage estimations, and did not handle object occlusions.

2.1.1.2 Probabilistic based techniques

Sayed Hossein *et al.* (2013) have presented a object tracking method that uses the spatio-temporal Markov random field (STMRF) model to track moving objects in H.264/AVC-compressed video sequences. This method combines the temporal and spatial features of the motion of the objects. In this method, the object tracking was performed through the use of block coding modes and motion vectors (MV) derived from the compressed bit stream. At first, the object to be tracked was selected manually in the first frame and the tracking was performed via successive frames of the video. This method evaluates the motion vectors of intra-coded blocks and global motion (GM) parameters in every frame of the video and then, separates the global motion parameters from the motion vector field. Hence, this method requires low processing time and provides correct results. Then, the target object was detected and tracked by Spatio-Temporal Markov Random Field model. The global motion parameters were used to find the object's position in the current frame by using the objects position in the previous frame. Then, the Iterated Conditional Modes (ICM) was used to update the position of the object depending on the temporal and spatial coherence. Even though this method provides the considerable performance, it was suffered from the problem of high computational complexity.

Serhan *et al.* (2016) have suggested an object tracking mechanism, that detect the object from a compressed video sequence through the use of H.265/HEVC standard. This framework was based on the motion vectors (MV) and block types. The bit stream of the video was decoded partially to get a block types which uses the domain information of the pixels to differentiate among the objects. This framework depends on the Markov Random Field (MRF) model and the MRF model was used to find the temporal and spatial information of the moving objects and it was updated in every frame of the video. This method follows the pixel domain method which was used to extract the colour information of the frames that are fully decoded and it was updated after completing the Group-of-Pictures (GOP). The experimentation was performed in a video sequences and the results of the experiments shows that this hybrid framework achieves higher performance in terms of accuracy. This framework was used to track objects from the videos with similar movement of multiple objects. The performance of this method was increased by combining the textual feature. Therefore, the correctness of this framework was improved in an effective manner.

2.1.1.3 Feature components based technique

To perform the Hammersmith pulled-to-sit examination, Debi *et al.* (2012) have presented a method in which follows the object tracking from video sequences. Hammersmith Infant Neurological Examination (HINE) is defined as the set of tests which are performed to evaluate the neurological development of babies who are at the age of 0-3 and the babies who are born after a pregnancy significantly shorter than normal, especially after no more than 36 weeks of pregnancy. These tests are carried out in hospitals to evaluate the development of babies with disabilities. In this circumstance, the authors have MOTSFK a video object tracking algorithm based on a node pruning and dynamic programming. At each step, feature point-based tracking algorithm chooses the best n nodes and the pruning algorithm was used to minimize the time required for tracking. At first, the feature points were identified in the initial frame of the input video and these feature points were tracked to find the trajectory of the object. This algorithm tracks the motions of the babies. Then, the pulled-to-sit event was examined by a geometrical model of the scene. This method was useful in hospitals to handle HINE tests. However, this method had limitation in the need of large field trial for validating the tool.

Kim *et al.* (2011) have MOTSFK an integrated shape and feature-based object tracking algorithm for tracking non-rigid objects. This algorithm was combined with an adaptive background generation which was used to avoid the problem of block matching. The background generation process was used to eliminate the blocks with moving objects which generates more matching errors. Median filtering method was used for recompense the blocking artefacts due to the eliminated blocks and blocks the performance of the moving objects. The background process was used to find the object's motion and the variation in shape. Then, the morphological edge operations were used to find the boundary of the objects. With the background information and the shape and boundary information of objects, this method provide better tracking performance since the target object and the feature points of the target object were found easily. The shape control points (SCPs) were allocated to the object's contour and the centroid was updated at the time of tracking in which the drifted SCPs are separated and the qualified SCPs were remains in the tracking process. The target object was differentiated from the background by the boundary information which classifies SCPs and the candidate of SCPs (CSCPs). The shape of the object was determined by the SCPs and SCPs need updating during the occlusion of objects which was performed by the CSCPs. Both the deformable objects and the generation of background were determined by the block matching method. During object occlusion, CSCPs took the place of the current SCPs which depends on moving region's size and the total SCPs. This algorithm was performed well even if unexpected movement of object occurs. This method did not affected by the failing factors like object occlusion, deformation of objects, and the spatio-temporal similarity among object and its background. The experimentation of this method was conducted in video sequences consists of different objects like walking people, swimming fish and so on and results obtained from the experimentation shows that this method performs well even object occlusion. The limitation with this method is that the tracking performance provided by this method was very low.

Chenglizhao *et al.* (2015) have suggested a rapid low-rank coherency analysis model to track objects from the video sequence. Initially, not only the local information and the global information were gathered from the successive frames of the video. Then, the low-rank coherency enables the tracking of object and took actions to prevent from object occlusion, illumination changes, unexpected movement and so on. Here, the global time-continuous target coherency and the features of local space-distinctive candidate were integrated with the low-rank coherency analysis. The compressive sensing based patch-level feature descriptor was used to represent the local candidate and this descriptor was obtained from the colour distribution of the frames. Then, the space-time feature set was generated by arranging the entire candidates in the frame catch and the frames to be processed. After generating the space-time feature set, the global coherency voting was enabled by the low-rank decomposition. This method tracks the moving object robustly even if sudden changes in the movements of the objects because the low-rank coherency inferred the co-occurring parts of various observations of the target. Then, the tracking results of the previous frames were combined with the low-rank estimation in the current frame. Hence, this method needs less time for perform computation and provide best performance. The experimentation was performed to evaluate the performance of this method regarding to correctness, robustness, and reliability and the results of experimentation shows that this method had good performance in above mentioned measures. The unexpected rotation of the target broke the low-rank coherency which creates some drawbacks to this method.

Shinsuke *et al.* (2016) have suggested a vision sensor system based on a scale-invariant feature transform (SIFT) algorithm. The MOTSFK vision sensor system uses two parallel processing methods. They are a resistive networks (RN) filter and digital processing blocks in a field programmable gate array (FPGA). The SIFT algorithm was performed based on two processes, such as whole-image parallel filtering and frequency-band parallel processing. The major components of the vision system are a metal-oxide semiconductor (MOS)-based resistive network, an active pixel sensor, a digital computer and a field programmable gate array. The metal-oxide semiconductor based resistive network was used to perform spatial filtering and provides a configurable filter size and the FPGA was used to performing the pipeline processing of the frequency-band signals. The evaluation of this method was performed by finding the feature points of object in a video frame. The advantages of this method are it consumes less power and require less processing time. The experimental results show that this method provides promising results to object tracking and detection. However, this method had some limitations like restricted number of scales and inaccurate SIFT algorithm.

Guang *et al.* (2016) have MOTSFK an algorithm named as local region sparse appearance model for tracking object from a video. This algorithm divides the object into number of sub-regions and performs clustering in every sub-region to acquire the sparse dictionaries. Hence, the object's spatial structure information was determined and the difference in appearance of the object was reduced. The object dictionary base was developed by integrating the dictionaries of all the sub-regions of object. Then, the space was defined among the various parts of the objects. At the same time, the operations like noise removal were performed to obtain worthy information. The update mechanism was performed by the template set updating mechanism. Here, the template set was filled by the object samples which are more valuable. The worthless object samples did not placed into the template set although the template set has space for putting object samples. The information of the time domain of the object was determined as weighted sum by using the patch sparse coefficient histogram. Thus, the better candidate object was acquired from the template basis. When the tracking method finds the location of the object incorrectly, this algorithm make use of dynamic sub-region resampling method which relies on cosine angle and find the correct position of the object. Hence, this algorithm provides guarantee for non-losing object. The experiments were conducted in a demanding video sequences and the results shows that this algorithm performs object tracking with considerable performance.

Guoheng *et al.* (2015) have suggested a non-rigid object tracking mechanism from a video sequence which was based on interactive segmentation and super pixel Gaussian kernel. The traditional object tracking methods use the bounding box to locate the target object but this method exploits the interactive segmentation which consists of the user-defined marker. The interactive segmentation was used for performing segmentation of objects in the initial frame of the input video sequence. This method prevents the background influence problem occurs in the bounding box. Then, the tracking was performed by the super pixel Gaussian kernel. The Gaussian kernel tracks the every pixel to determine the object in the next frame. The experimentation was performed on different video sequences and the results shows that this method obtains accuracy. The disadvantages of this method are, it did not provide robust tracking and did not handle object occlusion. The results of tracking was depends on the number of super pixels of the object.

2.1.1.4 Searching based technique

Badri *et al.* (2011) have suggested an algorithm for detection and tracking of moving objects from a video sequence which consists of two schemes. The first scheme was used to performing spatio-temporal spatial segmentation and the second scheme was used to performing temporal segmentation. A spatio-temporal spatial segmentation was performed by a compound Markov random field (MRF) model which considers the colour's spatial distribution, edge map in the temporal frames, and temporal colour coherence. Here, the segmentation process was reviewed as the pixel labelling problem which was resolved by the maximum posteriori probability (MAP) estimation technique. The combination of MRF and MAP framework was suffered from the problem of extensive computation which was brought by the random initialization. In order to avoid this problem, change information based heuristic initialization technique was MOTSFK in this paper. In this technique, the first frame was segmented by the compound MRF model and the estimation of MAP was performed by a hybrid algorithm which was formed by combining iterative conditional mode (ICM) algorithm and simulated annealing (SA). The first frame's segmentation result and the change information of the other frames were used to produce the segmentation of other frames. Then, the iterative conditional mode algorithm was performed on the frame which starts from the acquired initialization to perform segmentation. The temporal segmentation was performed by the change detection mask (CDM) depends on the label variation of two frames. Then, the object was tracked by the centroid based tracking technique. The advantage of this method was that it provides accurate tracking results. The disadvantage of this method is that it did not solve the object tracking problem in videos taken by the moving cameras.

Faegheh and Mohsen (2016) have MOTSFK a meta-heuristic approach for tracking objects from a video. In this method, the galaxy based search algorithm (GbSA) was modified. Then, the modified galaxy based search algorithm (MGbSA) was used to increase the correctness of the particle filtering framework. Initially, the particle filter uses the state transition model to generate the particles. After generating the particles, every particle was applied to the modified galaxy based search algorithm. The MGbSA was used to determine the optimal state of the object by calculating the fitness function through the likelihood model and simulate the movements of spiral galaxy to find the state space. After all the particles are processed on the MGbSA, the particle with high fitness value was taken as the state of the object. The experimentation of this method was performed in videos consists of changes in appearance, unexpected motion of objects, background clutter, and difference in pose and rotation to find the correctness. The results obtained from the experiments shows that this method provides robustness and correctness but it did not handle object occlusion. The possible future work of this method is to include the object occlusion handling mechanism to handle the occlusion of objects during the tracking process.

2.1.2 Multi-object Tracking Algorithm

In the multi-object tracking, the tracking algorithm tracks multiple target object from the video. The multiple object tracking can be applied for tracking the objects in the conference hall, traffic monitoring etc. The existing methods use the multi-object tracking. The multi-object tracking uses the following algorithms,

- Graph based techniques
- Learning based technique
- Feature components based technique
- Searching based technique
- Similarity function based technique
- Probabilistic based techniques

2.1.2.1 Graph based techniques

Amit Kumar *et al.* (2017) have MOTSFK the multi-object tracking (MOT) method based on sporadic appearance features. This method constructs the complementary graphs to determine the spatio-temporal information and object appearance information. Then, the multi-object tracking problem was considered as the consistent labelling problem. At each time stamp, the detections were found then, the aim of this algorithm was to connect the detections between times. This process was done by associating the labels on a graph sets, and every graph was used to find the process of assigning labels to the detection pair which was performed by the spatio-temporal information or the appearance information of the objects. The construction of the graphs was performed by the linear embedding of the features detected. In the appearance graph, the node's neighbourhood was defined to consider all the nodes. Therefore, the appearance features of this algorithm were present sporadically. After designing the graphs, the multiple object tracking was performed to find the assignment of labels which are consistent with every graph constraint resulting in the difference of convex (DC) program. In this method, the global objective function was divided into node-wise sub problems. This provides the efficient solution and constructs scalable and incremental graphs. Hence, this framework works on graphs with large size and practical tracking methods. This framework also supports the parallel implementation. The drawbacks with this framework are, it did not include the higher order motion model to construct the spatio-temporal graph and did not handle the confidence levels of features in a successive manner.

Longyin *et al.* (2016) MOTSFK a dense structure based multi-object tracking algorithm. Here, the undirected hyper graph was constructed depends on the tracklet affinity and the hierarchical dense structures was exploited on it. The hierarchical dense structures consist of group of nodes which were connected with the hyper-edges and have the affinity values. The similarities of the appearance and movement of tracklets between the spatio-temporal domains were examined through providing the higher-order similarities. To provide the efficient computation, a hypothesize-and-test algorithm was MOTSFK which was used to evaluate the hyper-graph and the dense structures were extracted from those graphs. This algorithm provides the better tracking results even long-term object occlusion. The experimentation was conducted in various challenging data sets with multi-pedestrian and multiple objects.

Yonatan *et al.* (2016), performed the multiple object tracking based on the dominant set clustering based tracker. The dominant set clustering based tracker was used to determine the dominant sets from the weighted graph. Most of the existing object tracking techniques considers only the few frames of the video but this method considers the relationship between both appearance and position detections among the entire video for performing the data association. The tracklets were found by the temporal sliding window and then merged. The merging of tracklets provides the ability to tracker for handling object occlusions and for providing robustness. The experimentation of this method was performed with the existing object tracking methods and results obtained from the experimentation shows the superior performance of this method. This method did not track the objects from the videos captured by the moving cameras and it was not suitable for multi-camera setup. This method only considers the people for tracking and it uses the minimum similarity measures.

Zhenyu *et al.* (2016) have suggested an object tracking mechanism based on the Connected Component Model which was used to globally solve the data association problem. The relationship among the observation association was defined as the equivalence relationship in the problem of data association. Depending on the spatial-temporal constraint the object's trajectories were did not joint. Hence, the Multi-Dimensional Assignment (MDA) problem was splitted into sub problems which are not dependent on each other via equivalence partitioning. The Connected Component Model (CCM) was used to solve the MDA problem, by providing the data association constraints and the equivalence relation. The CCM solves the data association problem in an efficient way and acquire the best results. The CCM was also used to solve the benchmark problems which occur during the tracking process. The CCM was used to solve the multiple objects tracking problem in an efficient manner and provide the optimal results. The experimentation was conducted on global data sets and the results obtained from the experiment shows that this algorithm provides better performance. The disadvantage of this algorithm is that the trajectory of the objects must be disjoint otherwise it provides the inaccurate results.

2.1.2.2 Learning based technique

Mohamed *et al.* (2016) have suggested a method for tracking multiple objects from a video sequence which depends on the sparse representation (SR). In this method, the object was recognized by the online dictionary learning method. Once the objects were detected, ever detected objects in the video was represented by the object descriptor which consists of both appearance and position features of the objects. The object which was detected is then subjected to the classification process and indexing process in accordance with the sparse solution which was acquired from the orthogonal matching pursuit (OMP) algorithm. In real-time applications, the tracking was needed to perform fastly, and the tracking results were did not affected by the inaccuracy. Here, not only the description of the objects and sparse representation was guaranteed to produce the exact identity of the objects which are moved in the video sequence. The OMP algorithm had high complexity on fewer steps and it was upgraded by the Graphics Processing Unit (GPU) implementation. Anyhow, after the process of object detection, the dictionary was enlarged and the descriptor size was also very large. Therefore, the process of system was decreased. To avoid this problem, here the OMP algorithm was implemented in GPU. The experimentation was conducted in various data set and the results of experimentation shows the efficiency of this algorithm and this algorithm requires less time for perform computation. This algorithm did not recognize the Multiview objects. This is one of the limitations of this algorithm.

Weiming *et al.* (2012) have MOTSFK multi-object tracking algorithm based on the learning approach. They have MOTSFK incremental log-Euclidean Riemannian subspace learning algorithm for tracking the target objects from the videos. Then MOTSFK algorithm utilizes the image features of the video with the covariance matrices. These features from the video are mapped with the vector space and the log-Euclidean Riemannian metric. The MOTSFK work is a based on the subspace learning algorithm. They have developed the log-Euclidean block-division appearance model for effectively analyzing the appearance of the each object in the video frame. The MOTSFK work finds the object appearance through the global and local spatial layout information of the object. This work uses the particle filtering based Bayesian state inference for the tracking of the multiple target objects from the videos. The use of particle filtering has effectively tracked the occlusion reasoning in the videos. The object appearance in the video gets changed in each frame. The change of the object appearance in the each video frame was tracked with the incremental updating of the log-Euclidean block-division appearance model. Due to the updation for the each change, the model construction takes more time for the tracking process. The major advantage of using this model is tracking of the objects can be done even if there is presence of occlusion in the video. The performance of the MOTSFK work is compared with the six state-of-art tracking algorithms. Experimental results show that the MOTSFK work has outperformed the existing works. This work has an improved tracking accuracy results than the other even at adverse conditions such as large variations in illumination, small objects, pose variations, occlusions, etc.

Rui Yao. (2015) MOTSFK a novel multi-object tracker based on online structured learning. This method was used to equalizing the structural classifiers of the object's training samples. Initially, the multiple object tracking was performed by the joint structured learning technique which makes the strong object's discrimination during the process of training and prevents the problem of the object occlusion and uncertainty of the identical objects. In the meantime, the joint structured learning model speeded up the training process and the non-linear kernels were then derived. The experimentation of this online multi-object tracking was performed in a challenging data sets and this method produces the correct and rapid performance for object tracking.

Shucheng *et al.* (2016) have presented a multiple hypothesis tracking method based on metric learning and multi-cue fusion (MHMHT) for tracking multiple objects from a video sequence. The appearance similarity of associations was computed by the distance among feature vectors which are presented in a tracklet and the track hypothesis templates. Then, the Kalman filter was used to fuse the appearance similarity with the dynamic similarity. The discriminative appearance metric was learned by the spatial-temporal relation of tracklets which were presented in the temporal window to make the discriminative appearance similarity and the discriminative appearance metric was used to calculate the distance among the salient templates and feature vectors. Then, the salient templates were updated by the incremental clustering method which considers the high order temporal context information. The experimentation of MHMHT tracker was performed on four benchmark datasets. From the experiment results, we understood that the MHMHT tracker out performs the existing methods.

Dorra and Guillaume-Alexandre (2016) have MOTSFK the online multi-object tracking method which depends on the multiple features of the objects. This method was used to track multiple objects even the object occlusion occurs. Here, the appearance model representation was used to choose the better location of the object. The target's appearance model was designed with the following features: a sparse appearance model, a colour model, a spatial information model, and a motion model. To choose the best target of the candidate, this method uses the linear affinity function which combines the similarity scores of every feature. In this method, the object tracking was viewed as a data association problem among the detection sets and target sets in accordance with the joint probability values. The feature descriptors were combined with the data association method in which all the targets and candidates are matched with each other under geometric conditions. This method was able to handle the object occlusion problem by the hierarchical data association process where the target objects were decomposed into occluded and unoccluded targets. The quality of the detection response was increased by the position interpolation of the lost target. The experimentation was performed on the challenging public video sequences and the experimental results show the robust performance of this method and the method had the ability to handle the different challenging problems, such as object occlusion, The hybrid tracking method has been evaluated on public video sequences. Compared with lose similarity among the appearance models of the target objects.

2.1.2.3 Feature components based technique

The scale and rotation invariant object descriptor and detector were MOTSFK by Abhineet *et al.*(2016), which were based on the Speeded Up Robust Features (SURF) algorithm. In this method, the objects were detected in the current frame by doing comparison with previous states of that objects in the previous frame. The object detector was used to find the object similarities by analysing the characteristics of that objects and calculate the centroid of the particular object in each frame of the video and specifies the object's trajectory from the past frames of the video sequence. This method had the ability to track multi-objects from the single video. This system uses the Graphical user interface which makes the interaction between users and the system. The users have drawn the bounding box on the objects to be tracked and choose the initial frame. The experimentation was performed on MATLAB and this method tracks multiple objects in a more effective way. The feature tracking was performed by the SURF algorithm. At first, the feature tracking mechanism was performed on images and then performed on the videos. The results have been produced as a Graphical User Interface. The target objects were selected from the initial frame and the tracking was continued through the successive frames of the video.

Swati *et al.* (2016) have MOTSFK a multi-object tracking using TLD background. The MOTSFK work based on the TLD background has the better usage for the continuous tracking of object in a video stream. In this work, each object in the frame is modelled with the specified location and extent in the video frame. So, for the each frame the MOTSFK work finds the location and extent of object. This indicates that presence of the target object in the each frame. The MOTSFK modified template matching algorithm based on SURF algorithm and squared difference error method performs the multi-object tracking of the videos in the real time. In this work, the template matching of the objects depends on the various image features of the target object. SURF algorithm depends on the feature point detection for performing the template matching. They have utilized the template tracking algorithm for performing the multiple object tracking. This method utilizes the region of interest (ROI) from the object from a video stream from trained object database. The tracking is performed by the MOTSFK matching feature with the principle component analysis. The model cannot modify the weights of the algorithm due to the various influencing factors while tracking. The classification of the target objects has lesser accuracy.

2.1.2.4 Searching based technique

Beyan*et al.* (2012) have presented a multiple object tracking that works depends on the mean-shift algorithm. The object's foreground was detected by the Gaussian mixture which follows the removal of shadow and noise from the video. The kernel mask was used to reduce the search area of the objects. Detection of foreground could be useful to fine the objects which are leaving and entering into the search area of the scene. The object trackers were refreshed to avoid the problems that occur during the tracking process, like object occlusion, sudden changes in the movement of objects and so on. The shadow removal method was used to improve the correctness of the tracking process. This method provides the better tracking results for video captured by the static cameras. The advantages of this algorithm are that it was implemented in an easy manner and provide robust tracking results. This algorithm did not perform object tracking from the videos captured by the moving cameras.

Kwak *et al.* (2014) have MOTSFK the multiobject tracking algorithms based on the object segmentation and the data association. The data association is performed based on the integer programming and the object segmentation was done based on the boundary detection. The MOTSFK multisensor fusion-based unknown object detection and tracking approach allows the multiobject tracking in the videos. They segment objects in the each frame through the detection of the boundaries of the each object in their respective frames. Data association allows the tracking of the multi-object clusters by associating the data in the each frame. The MOTSFK algorithm combines the data segmentation and the association through the single-line scanning lidar and a camera. The real datasets obtained from a moving platform were used for analyzes of the MOTSFK work. The experimental results shows that the MOTSFK work for the object boundary detection and data association has better usage for the of sensor fusion.

2.1.2.5 Similarity function based technique

Hyunguk *et al.* (2016) have MOTSFK a data association method for video tracking with the multiple objects. The tracking was performed in a non-overlapping camera network. The MOTSFK data association method uses a similarity function with the multi-camera topology. This will effectively tracks the objects with the colour identification. The local tracking model tracks the objects through the person re-identification and color appearances of objects. The multi-camera environment has many challenges and hence the use of the re-identification is not enough to recognize target objects in real time scenario. Hence, they have utilized the data association techniques to extract the required features in a low resolution camera for the feature extraction process. They have developed the automatic thresholds values for the evaluation of the similarity function for the tracking process. From the simulation results it is evident that the MOTSFK work has better tracking performance than the existing methods for the non-overlapping camera network.

The difficult task in the multi-object tracking method was the how to combine the frame-by-frame detections. To avoid this problem, Jerome *et al.* (2011) have MOTSFK a technique in which the target motions were formalized as the flows in the graph edges. At first, this method uses the general object tracking framework in which the input was the occupancy map which was generated by the detector. Then, this method uses the k-shortest path algorithm to provide the optimal tracking results and minimize the computational complexity. This algorithm performs multiple objects tracking in a real time computer and it had the advantage of robustness. The advantages of this algorithm are: it was simpler to perform object tracking and provides the global optimal result of tracking. This method requires minimum amount of signal and small number of meta-parameters. However, this algorithm needs additional signal to perform tracking.

Mohamed A. N.*et al.* (2017) have suggested a collaborative model for multi-object tracking problem which was used to improve the interaction among the pre-trained object detector and single object tracker within the particle filtering framework. This method creates the relation among the object detections and tracking. Then, every detected region of the image was considered as a key sample to perform the online updating. In this method, the motion model was used to combine the object detection with the dynamics of the objects. The sparse representation and 2DPCA was used to generate the diverse features that increase the illumination changes between the object trackers. Then, the tracker's appearance model was updated by the sample selection method. The likelihood function and the association of data were performed by the discriminative model and generative appearance models respectively. The experimentation was performed on various videos and the results of experiments show the superior performance of the MOTSFK multi-object tracking method.

Shu *et al.* (2016) have introduced an online inter-feedback information among the object detection and tracking processes in the object tracking method. This tracking algorithm consists of two components. The first one was the object detection by feedback from the tracking process and the second one was the object tracking based on object detection. At the first step, the detectors were used to detect the objects by using the information gathered from the tracking process. The tracking was performed based on the information from the object detection. Here, number of tracking methods were used for track the multiple objects. There were two types of objects: single objects and multiple objects. Both single objects and multiple objects were dealt various strategies. The experiment was conducted in number of real time videos and this algorithm was acquired a better performance for tracking multiple objects. This algorithm requires minimum number of detectors and search scale. The similarity calculation of this algorithm was required improvements and the modification was needed to the flexible detector.

Portela Sotelo *et al.* (2012) have MOTSFK a multiple objects tracking problem as the problem of finding a patient and the equipment in a radiotherapy treatment room. This method uses the information, such as geometric information, visual information, and sematic information. This method generates the virtual environment by using the above information and the generic model. The images in the video capture by the different cameras were confronts the numerical representation of the scene. This method was also used to generate the control system in radiotherapy in which one can communicate with the person inside the treatment room. The equipments were integrated to the treatment room of the hospitals in every day and these equipments need to understand the position of the patients in the treatment room. This is the challenging task in medical applications. Hence, this method was useful for such a situation in hospitals. The experiments were conducted during no treatments. This method did not produce the absolute results which was the drawback of this method.

2.1.2.6 Probabilistic based techniques

Hidetomo (2013) has MOTSFK the tracking model for the multiobject tracking algorithm through the tracking and the learning process. They have MOTSFK the extended MCMC method for the tracking purposed and an extended HMM method for learning movement of the target objects in the videos. This work proposes a GUI based model for the tracking which utilizes the enhanced usability. In the MOTSFK work, a cost reduction method for the MCMC has been utilized based on the iteration loop of the Markov chain process. This work incorporates the stochastic model parameters for the stable and robust object tracking. The MOTSFK HMM method integrates various modules to deal with the multiple discontinuous trajectories present in the video. The MOTSFK interactive system allows the auto-allocation of symbols from images and a hand-drawing function for learning trajectories of the videos. The MOTSFK model has better algorithms for the spatio-temporal smooth and discontinuous trajectories.

In this chapter, various existing methods for the single object and the multiple object tracking were discussed. This chapter surveys a total of 37 existing methods for the object tracking from the videos. Various challenges and the methods used in the each technique have been discussed. The single tracking algorithm utilizes Filtering based techniques, Probabilistic based techniques, Feature components based technique, and Searching based techniques. The multi-object tracking models have utilized Graph based techniques, Learning based technique, Feature components based technique, Searching based technique, Similarity function based technique, and Probabilistic based techniques. Each model has its own advantages and disadvantages. From the above survey bit is concluded that the feature extraction without the object occlusion and the spatial and the visual tracking models are better suitable for the multi-object tracking.

2.2 LACUNA IDENTIFIED

Object tracking is the important problem in the video analysis. One of the major problems in object tracking is the difference appearance of the object and its surroundings. The target appearance will change during the prolonged tracking process because of differences in appearance or viewpoints of the objects. Although, the video is taken by the stationary cameras the background of the image will change. Object occlusion is another important challenge in the field of object tracking in which only the regions of occluded objects are able to seen and the similarities among the objects and their features begin to be equivocal.

CHAPTER 3

OBJECTIVE

The main objective of research is enlisted as follows;

- The primary objective of this work is to propose a Wavelet based object tracking model. The Wavelet based object tracking model performs object segmentation through shot segmentation. Various features such as shape and histogram based features are extracted during the feature extraction process.

- The secondary objective of this research is to propose the Kalman filter based multi-object tracking model. This extracts features such as the centroid and the tilt using shot segmentation and manual segmentation. The Kalman filter tracks multiple target objects from the videos with the use of the extracted features.

- The final contribution of this work is to propose a hybrid tracking model based on spatial and visual tracking. The spatial tracking in the MOTSFK model is based on the tangential weighted function, and the visual tracking model uses the second derivative based visibility model. The MOTSFK hybrid tracking model effectively combines the spatial and the visual tracking models and thus performs multi-object tracking.

CHAPTER 4

MATERIALS AND METHODS

4.1 MATERIALS

The application is performed with the help of MATLAB 14a, which was simulated in a scheme having system specifications of 6 GB RAM and 3.2 GHz Intel i-7 processor.

4.2 METHODS

The various methods used in the detection of multi-objects are as mentioned below. The main research area in the field of digital image processing and recognition is the detection of a moving target. The Detection of moving targets finds valuable applications in the field of robot navigation, intelligent visual surveillance systems, medical image analysis, industrial inspection and video image analysis. Advancement in technology has contributed towards the remarkable growth in the field of computer vision and technology, and object tracking gains remarkable interest in tracking the objects present in the video through tracking their motion in different frames and finding their location. These advancing technologies in the multimedia have produced a revolution in the field of object tracking. Even though a lot of techniques are used for tracking the objects, they failed to track multiple objects simultaneously.

Video surveillance plays an important role in traffic control, medical imaging, and various security applications. Video surveillance allows the tracking of objects from various frames of a video. The tracking model finds the exact location of the objects in the successive video frames. The basic issue behind the detection of the object in static images is the difficulty to categorize all the objects in the image into different classes, for which a robust object detection method is essential. The major function of the robust object detection method is to differentiate all the images even under uneven illumination, images that are rotated in the plane of the image, occluded objects, or the images that are blended in their background. For the detection of moving objects, the features corresponding to the higher density cause the rate of detecting the objects to have a very low value and good localization accuracy. For the features that possess high reliability, object detection yields low false detection rates. The noise in the detection process may be due to the real world environment and also due to the problems existing in the video acquisition systems and the transmission channels. These noises directly cause poor video quality mainly, for the high-degree of vision. Hence, to perform video tracking, the frames in the video are required to locate and track the object in the frame. A lot of algorithms are employed for tracking the objects. Video frames provide more valuable information about the object when compared to still images, as they are changing with respect to time. The tracking algorithms used in the existing systems faced a lot of disadvantages like non-homogeneous background, high cost, inability to handle low clarity images, and to deal with the varying backgrounds that change the scale, pose and look of the image. Therefore, for effective tracking, the algorithm used should be adaptive, robust even under a noisy environment, and applicable in real-world scenarios.

The vast technical advances and the widespread development of the multimedia have contributed towards the development of an effective processing method that handles the huge database. The efficiency of the processing mechanism relies on the effective approach used for the video processing system. Therefore, to enhance the efficiency of the processing system, an effective search mechanism is required. Therefore, this chapter focuses on an effective method for performing video processing which is analysed using three methods. The essential requirements for effective processing of the video from the huge mass of the video archives include video object detection and the tracking system. The video contains multiple objects with various features. The objects in the video get classified in two ways. They are, i) objects in the videos behaving independently, and ii) objects in the videos which are dependent on each other. The Video of the ship moving in the sea is categorized as an independent object. But, the video of a person moving in the crowd or a cinema hall can be categorized as a dependent object. Tracking of the dependent object in the successive frames of the videos is a challenging task. The objects in the multiple object video have a small size and similar features. Due to this, the accuracy of the tracking process gets reduced. The tracking model faces difficulties when the target objects change their motion along the subsequent frames. The tracking model in the video surveillance is categorized in two ways. They are,

- Spatial tracking
- Visual tracking

The visual tracking model predicts the change occurring in the tracking object for each frame of the video. For example, the appearance of the target objects in a highly crowded region for each frame of the video. The spatial tracking model detects the location of the tracking object in each frame. The object detection and the tracking system utilize the approaches, namely, shot segmentation and feature extraction approaches. In the process of object detection, the videos are converted into a number of frames using the approach named as shot segmentation. In the shot segmentation process, the videos are initially converted into shots and the shots are represented as blocks. Using the distance measure, the distance between the frames is calculated and the frames in the video depend on the minimum distance measure. All the frames in the video undergo the selection process using the distance measure. Once the frames in the video are selected, the background separation process is carried out to detect the object from the frame. The main criterion in the background separation method is the area constraint. The location of the object that satisfies the minimal area constraint is fixed and the corresponding object is noted for locating the object in the successive frames. The dimensions of the located object is determined and cropped. This cropped object acts as the reference object for cropping the object from the successive frames.

The second frame is checked to have the location of the object same as the location of the object in the first frame. If both the locations of the frame remain the same, then keep the location as the same. If the locations vary, then extend the dimensions of the second frame from all its edges, so that a new image is obtained. Then, crop the image accordingly. The same procedure is repeated for the other frames in the video. The location of the object in the frames is stored, as it is essential for tracking the object from the video. This chapter presents a brief description of the object detection and the object tracking processes.

The method used in this chapter describes the efficiency and accuracy of the tracking process and finds application in various video retrieval processes. This chapter also discusses the MOTSFK that possesses the capacity to track the objects in an efficient way that enables multi-object tracking. This method is more robust when compared with the other existing methods. The existing methods failed to track the objects efficiently, as they obey the background-based object tracking principle. On the other hand, the most commonly employed tracking method is the moving target detection, which has been employed in most of the target areas. Even though moving object detection is used in most of the applications, it faces a lot of challenges like complexity and other external factors. Therefore, to overcome all the issues, a robust tracking method is required. The previous chapter dealt with single object tracking and this chapter uses a technique for multiple object tracking and detection, which improves the robustness of the detection technique. This multiple object detection technique uses the shape-based features and the Kalman filter. At first, the video is segmented into a number of frames and the first two frames are initially considered for detecting the objects. The distance measure is introduced that is used to locate the significant objects in the frame. The significant objects selected will possess a minimum distance measure. Then, the area constraint is used to fix the object in the frame exactly and the object with the minimum area is selected. In other words, the object whose area is less than the threshold area is selected. For the selected object, determine the corners, centers, tilt, height, and width of the object. Finally, the Kalman filter is used to determine the centroid of the object. If the newly determined centroid, determined using the Kalman filter, is the same as the old centroid of the object, then it determines the position of the object or else, the Kalman filter re-computes the centroid, which is the object tracking phase. The performance analysis of the MOTSFK carried out in terms of two parameters, namely, the error value and the score value, proves the effectiveness of this method.

This research work proposes a hybrid tracking model to overcome the disadvantages of the previous models. Hybrid tracking performs the multi-object tracking through the combination of the visual tracking and the spatial tracking approaches. Hybrid tracking uses spatial selective selection for object tracking in the video with high-density objects. This research work discusses the various challenges in the multi-object video, to effectively build the tracking model. The video with the high-density object suffers from object occlusion. Object occlusion occurs when the target object gets hidden by the other objects in the frame. The background of the video plays a major role in the object tracking process. The video of the moving people in the snowy environment makes the tracking process difficult. The higher density crowd environment has the rapid movement of the people and they appear small in size.

This research work overcomes the challenges in the multi-object videos. The research work incorporates both the visual and the spatial tracking models for the efficient location of the objects in the video frame. The multi-object tracking performance gets reduced due to the inter-object occlusion, object confusion, different posing of objects in the frame, environment with heavy clutter, small size of the objects, similar appearance of the objects, and interaction among the multiple objects. The research work proposes a hybrid method for effective multi-object tracking in the video. This is the combination of the visual and the spatial tracking models. The visual tracking model is based on the second derivative function. The spatial tracking model performs the tracking with the use of the tangential weighted function. The steps involved in this research work are explained as follows;

- The primary step involves finding the objects in the video through the k-means clustering process.
- The multi object tracking involves the visual tracking model based on the second derivative function to predict the objects in the videos.
- The second step performs the spatial tracking with the tangential weighted function.

The hybrid model MOTSFK in this research work does the above steps to track the multiple objects in the videos effectively. The second order derivative function has the maximum flexibility since it uses the Neighbourhood algorithm for the search process. The simulation results are obtained by implementing the model for tracking the videos from the UCSD dataset.

4.2.1 Visual Tracking By Wavelet Moment, Shape and Histogram For Single Objects

The widespread expansion in the field of multimedia has increased the necessity for an effective method for detecting and tracking the objects, as the existing methods faced a lot of challenges including the long tracking time as a result of the inability to converge to an exact location of the image. In addition, it also suffers from errors that affect the proper matching of the image, inability to deal with the tracking of the objects in the varying environment, and failure to track the color objects. To overcome the issues mentioned above, this method is an effective one. An effective method is needed to handle operations like the retrieval and processing of videos from huge databases. The effectiveness of the retrieval and processing methods mainly depends on the techniques employed for tracking and searching.

This method employs an effective technique for the object detection and tracking system. The MOTSFK technique uses the segmentation technique named as the background subtraction based segmentation and wavelet moment for segmenting the videos, and the process of segmentation enables object detection in the video, and the features used in this method depend on the shape and histogram-based feature extraction techniques. Hence, visual tracking consists of two modules for object detection and object tracking (Fig :1).

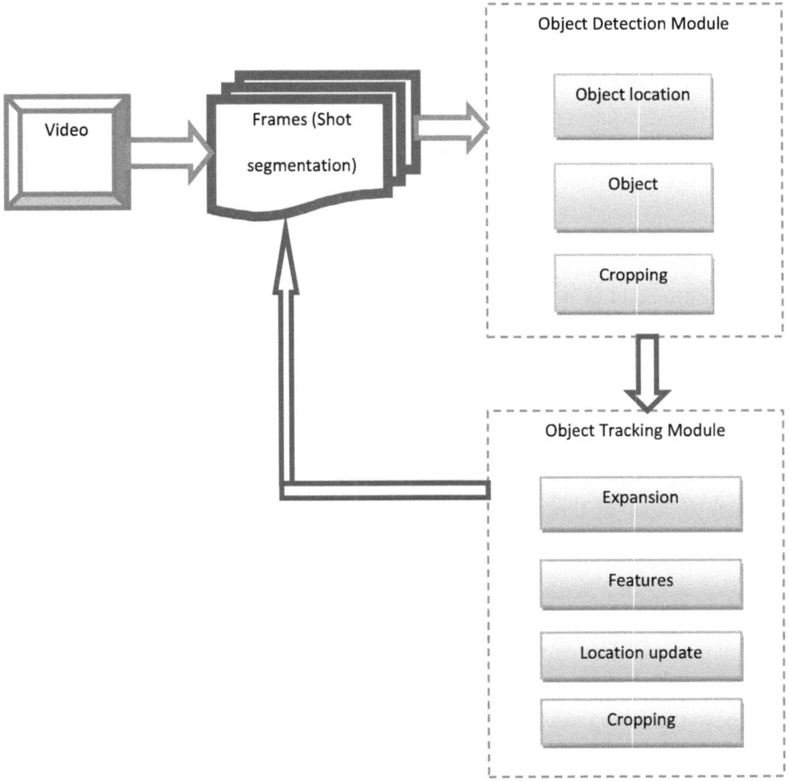

Fig 1 : Block Diagram of The MOTSFK Visual Tracking Method Using Wavelet Moment, Shape And Histogram Features

4.2.1.1 Object Detection Module

The Object detection module is the module that detects the object from the frames. In this module, initially, the input video is divided into frames by a technique named as the shot segmentation technique. Then, the object in the frame is detected using a technique called as background subtraction. Once the object in the frame is determined, the parameters of the object are determined. Then, the required object is cropped from the frame.

Shot segmentation

Shot segmentation is the most important step in the object detection phase. Shot segmentation is the first step and is used in most of the applications like video retrieval, video watermarking, and video summarization. The main process involved in the shot segmentation process is the segmentation of the videos. Videos considered for object detection are divided into non-overlapping units. Each of these non-overlapping units is called shots. On each of these shots, one can find a considerable difference from other shots, and this change is mainly due to the background noise and as a result of the objects present in the video. For performing the shot segmentation of the videos there are a lot of techniques.

Among all the existing techniques present in the literature, this paper uses the shot segmentation technique that is based on the wavelet transform. Initially, the videos that are converted into shots are presented for shot segmentation using the wavelet transform-based shot segmentation. The word wavelet is nothing but the mathematical function whose average value remains zero in a certain allowable interval. The selection of the number of frames from each shot depends on the distance measure and the transformation. At first, two frames forming a shot are selected and they are represented as blocks. Once the block representation is made, then the DWT is applied to all the individual blocks of a frame. For illustration, let us consider that there are n number of real numbers that are the inputs to the segmentation as represented below,

$$x = x_1, x_2, \ldots \ldots x_n \qquad (4.1)$$

where, n is the total number of inputs, x_n is the nth input. Now, the DWT representation of the k sequences of the z real numbers is given as,

$$\varpi_{kz} = \int_{-\infty}^{+\infty} x(\tau)\psi_{kz}(\tau)\,d\tau \qquad (4.2)$$

where, $$\psi_{kz}(\tau) = \frac{1}{\sqrt{\alpha_{jk}}}\psi\left(\frac{\tau - \beta_{kz}}{a_{kz}}\right) \qquad (4.3)$$

ψ is the mother wavelet, α is the scaling parameter, and β is the shifting parameter respectively. Now, the distance measure is calculated, which is the distance between the frames. The distance measure is denoted as Δ_{ij}. Let us consider the two frames Γ_i and Γ_j. Now, the distance measure is calculated based on the following formula.

$$\Delta_{ij} = \sqrt{\sum_{j=1}^{m}\left(\Gamma_i - \Gamma_j\right)^2} \qquad (4.4)$$

where, Γ_i and Γ_j are the two frames considered for the distance measure, and Δ_{ij} is the distance measure. The distance measure is calculated for the two selected frames and the same formula is used for calculating the distance measure of the other consecutive frames. As is clearly shown from the above explanations, the number of frames selected from each shot depends on the distance measure, the distance measure is calculated and the frames are selected based on the minimum distance between the frames. The minimum value of the distance measure indicates that there is maximum similarity existing between the two frames. Let us consider the sequences in the input video as $G(j,k)$ and the non-overlapping shots determined using the shot segmentation process denoted as $S(j,k)$.

Background separation

For performing the background subtraction process, at first, consider the first frame from the shot for detecting the object in the corresponding frame. The method employed for detecting the object from the frame is the mode-based background separation method. The Mode of the image location is nothing but the pixel value of the most often occurring image at the defined location. Let us consider that the location of the image as lo_i, and let us consider that the frequently occurring image at the location lo_i is represented as $\{p_1, p_2, \ldots p_k\}$ and the frequency of these pixels is given by $\{f_1, f_2, \ldots f_k\}$. Now, the mode of the location lo_i is represented as $\mathrm{mod}(lo_i)$. The equation of $\mathrm{mod}(lo_i)$ is,

$$\mathrm{mod}(lo_i) = p_j \qquad (4.5)$$

where f_j is the maximum among f_i with $1 \leq i \leq k$.

These pixels obtained using the mode function are termed as mode pixels and they are considered as the background pixels. These background pixels do not undergo motion and hence, the color of these background pixels remains the same when the process is in progress. But, for the foreground pixels, the object is moving and hence, the pixels of the object undergo color changes, and the color of those foreground pixels acquires different color and shape with respect to the time. Due to this reason, in this chapter, an assumption is made. The assumption is that the most often occurring color pixel vector in the frame is the background pixel color vector. Therefore, the pixels that do not come under the background pixels are considered as the object pixels and the location of the object pixels is determined. For example, if we consider, m number of background pixels present in an image location, the representation of the background pixels is given as,

$$B = \{b_1, b_2, \ldots \ldots, b_m\} \qquad (4.6)$$

Now, the object pixels in the location are given as, $\{O\} = \{P\} - \{B\}$. The objects in the frame are represented as, $O = \{o_1, o_2, \ldots \ldots, o_{n-m}\}$. Now, the pixel location of the objects mentioned above is, $\{LO\} = \{lo_1, lo_2, \ldots \ldots lo_{n-m}\}$. Once the location of the object is computed, the dimensions of the object are computed, which is essential for appropriate cropping. The dimensions of the object include the centre of the object, width of the object and the height of the object. To determine the dimension of the object, at first, fix the area constraint and determine the object based on the minimum area. The purpose of using the area constraint is that some of the object pixels that are detected may contain noise; as a result of the noise, the real pixels of the object are not presented. Therefore, to avoid noise, the object with the minimum area is presented. The minimum area, which is the threshold area, is denoted as Th_a. The area of the object should lie below this threshold value. Let o_i be the object pixel and A_i denote the area of the object pixel o_i, then the area constraint is given by the following condition.

$$if \quad A_i < Th_a, \qquad then\, discard\, o_i \qquad (4.7)$$

Once the object pixel is determined using the area constraint, then determine the dimension of the object pixel. Let us consider the dimensions of the object o_i as $E_R(i), E_L(i), E_T(i),$ and $E_B(i)$.

where, $E_R(i)$ - right edge of the object image o_i.

$E_L(i)$ - Left edge of the object image o_i.

$E_T(i)$ - Top edge of the object image o_i.

$E_B(i)$ - Bottom edge of the object image o_i.

Using the four object edges $E_R(i), E_L(i), E_T(i),$ and $E_B(i)$, the object centre of the ith object is computed.

$$OC_x(i) = \frac{|E_R(i) - E_L(i)|}{2} \quad and \quad OC_y(i) = \frac{|E_T(i) - E_B(i)|}{2} \qquad (4.8)$$

where, $OC_x(i)$ and $OC_y(i)$ are the object centres i and j of the object image o_i. The width and the height of the object image o_i are calculated using the following formula.

$$OW(i) = |E_R(i) - E_L(i)| \qquad (4.9)$$

$$OH(i) = |E_T(i) - E_B(i)| \qquad (4.10)$$

where, $OW(i)$- Width of the ith object image.

$OH(i)$ - Height of the ith object image.

These calculated dimensions of the image are required for cropping the image to determine the required object. In simple terms, cropping is defined as the process of cutting the image into a smaller size such that the cropping process discards the irrelevant or the less important parts of the image. In other words, cropping ensures to extract the most important part of the image, and the cropped image is termed as the object. Finally, resizing is done that resizes the object, which fixes the size of the object itself.

4.2.1.2 Object Tracking Module

The Object tracking module tracks the positions of the detected object present in the successive frames. Initially, the first frame is chosen for detecting the objects of the frame, and finally the other successive frames are considered for detecting the objects. The dimension of the detected objects is determined in the previous section. Once the dimension of the first frame is determined, and then the second frame is cropped based on the first frame.

Let $OC_x(i), OC_y(i), OW(i)$, and $OH(i)$ denote the location of the object in the frame and after determining these parameters for the first frame, the parameters of the second frame are determined. Now, the cropped image of the first frame is represented as C_1 and the cropped image of the second frame is denoted as C_2. The main concept is that if the object cropped from both the frames is the same, then keep the location of the object the same. The condition is given as,

$$if \ (C_1 = C_2), \quad keep \ the \ object \ location \qquad (4.11)$$

If the cropped objects from both the first and the second frames are different, then the dimension of the cropped object is enlarged from all the sides. The dimension of the image $E_R(i), E_L(i), E_T(i),$ and $E_B(i)$ is extended from all sides to obtain a new image with new dimensions like $Ex_R(i), Ex_L(i),$ $Ex_T(i),$ and $Ex_B(i)$. For this newly obtained image, determine the features like density, correlogram, histogram and wavelet moment.

The term density of the newly updated cropped image is defined as the total number of pixels present in the cropped image per unit measure and it is represented in pixels per inch (PPI) or represented as pixels per centimeter (PPCM). Let us consider that the cropped image consists of y number of pixels and z is the measure in centimeter;, D is the PPCM density, then the PPCM density is represented as shown below,

$$D = \frac{y}{z} \qquad (4.12)$$

The term correlogram is the term that denotes the autocorrelation of the cropped image values. In other terms, correlogram is the cross-correlation of itself. Now, the word auto-correlation is explained through a formula given below,

$$g(x,y) * g(x,y) = \int\limits_{-\infty}^{\infty}\int\limits_{-\infty}^{\infty} g(x',y') g(x+x', y+y') \, dx' dy' \qquad (4.13)$$

where, $g(x,y)$ - Two-dimensional functional that represents the input image with dimensions x and y respectively.

$g(x',y')$ - The dummy variables of integration.

The third most important feature extracted from the newly obtained cropped image is the histogram. The Histogram represents all the pixels of the newly obtained cropped image graphically. It represents all the frequencies that are distributed over the discrete intervals with their area proportional to the frequency of the pixel in the cropped image.

An important feature extraction method to extract features from the cropped image is the image moments method that generates high discriminative features. The Wavelet moment belongs to the family of the orthogonal moment family, which uses an orthogonal wavelet function known as the kernel. The wavelet moments combine the advantageous features of the wavelet analysis and the moment analysis, and create the improved moment descriptors. The Following is the formula that shows the wavelet moment of the input image $g(x, y)$ with the dimensional area $m \times m$.

$$W_{abq} = \sum_{x=1}^{m} \sum_{y=1}^{m} \psi_{ab}(r) e^{-jq\varpi} g(r, \varpi) \qquad (4.14)$$

$$\text{where, } r = \sqrt{x^2 + y^2} , \; \varpi = \tan arc\left(\frac{y}{x}\right) \qquad (4.15)$$

$\psi_{ab}(r)$- Basics of the mother wavelet.

Thus, all the four features like density, Correlogram, histogram and wavelet moments are extracted for (or 'from'?) the enlarged image, which is the newly obtained cropped image. The four features extracted from the newly obtained cropped image are represented as $den_i, cor_i, hist_i$, and $wavm_i$. Among all the features generated for the newly obtained cropped image, select the location of the feature of density, histogram, and wavelet moments with the higher values of the density, histogram, and wavelet moment features obtained and the correlation feature selection is based on the lower values of the correlation features obtained. If the location of the four features is the same, then fix the location of the object and use this location and the object parameters to crop the object in the second frame.

$$if \ \left(Lo\left(\max den_i\right) = Lo\left(\min cor_i\right) = Lo\left(\max hist_i\right) = Lo\left(\max wavm_i\right)\right),$$
$$then \ take \ the \ position \tag{4.16}$$

where, Lo denotes the location of the object, max refers to the maximum values of the features, and it is used to denote the location of the features based on the maximum values of the features like density, histogram, and wavelet moments, and min denotes the minimum value of the features that determine the location of the object.

The above-explained process is repeated for all the other frames and the object in the subsequent frames is located based on the location of the object in the first frame.

In general, the object image of the $(i+1)^{th}$ frame depends on the image cropped from the ith object image cropped from the frame. The object tracking is based on the location of the object image in all the frames. Therefore, the location of the frame is saved to perform effective tracking.

1	**BEGIN**
2	Shot segmentation to extract frames from the video.
3	Read the frame
4	Determine the location of the object using the object subtraction method as $\{LO\} = \{lo_1, lo_2, \ldots \ldots lo_{n-m}\}$.
5	Calculate the dimensions of the detected object $E_R(i), E_L(i), E_T(i)$, and $E_B(i)$.
6	Crop the object
7	$if \ (C_1 = C_2)$, keep the object location
8	Else
9	Enlarge the cropped image on all sides to obtain new dimensions $Ex_R(i)$, $Ex_L(i), Ex_T(i)$, and $Ex_B(i)$.
10	Extract the features of the cropped image

11	*if* $\left(Lo\left(\max den_i\right) = Lo\left(\min cor_i\right) = Lo\left(\max hist_i\right) = Lo\left(\max wavm_i\right)\right),$ *then take the position*
12	Keep the record of the location of the object for object tracking.

Fig 2 : Pseudo Code For Visual Tracking Using The Wavelet Moment, Shape
And Histogram Features

The pseudo code of the research work presented in this chapter, depicts the visual tracking that uses the wavelet moment, shape and histogram features (Fig:2). Initially, the video undergoes a process named as shot segmentation for separating the frames from the video. Once the frames are obtained using the shot segmentation process, the object in the frame is located using a method known as object subtraction.. In the object subtraction method, the location of the object that occurs frequently is located (or 'saved' why use the word 'locate' twice). After determining the location of the pixels of the corresponding object, the object is subjected to verification as to whether it satisfies the area constraint. Area constraint is nothing but the threshold area that should be satisfied by all the objects in the frame. The concept behind the area constraint is that the object less than the area constraint will be selected, which means that the object contains the maximum similarity. The objects that satisfied the area constraint are selected.

The dimensions of the detected object in the frame are detected and are computed. The dimensions include the centre point of the object in the frame, and the height and width of the object that depend on the edges of the object. The height of the object depends on the top and bottom edges of the object in the frame and the width of the object depends on the right and left edges of the object in the frame. Once the dimensions of the detected object from the first frame are determined, the same process is repeated for the other frames for determining the object. The object in the first frame is cropped and it is used as the reference to crop the objects in the other frames. If the location of the cropped object in the two frames is the same, the location of the object is set to remain the same. When the location of the cropped image obtained from the second frame varies from the location of the reference object, then extend the dimensions of the newly obtained cropped image in all its directions. The same process is repeated for all the successive frames. The location, and dimensions of the objects in the frames are stored so that it is necessary to track the object in the video.

4.2.1.3 Experimental Results

This section depicts the result of the analysis of the visual tracking method in detecting and tracking the object, which involves the database used for tracking the object with the experimental setup and the evaluation metrics used for evaluating the performance of this method. This section presents the screen shots of the implementation process and gives a detailed description of the performance analysis.

4.2.1.3.1 Database Employed

For evaluating the performance of the visual tracking method of object detection and object tracking, the database is used that involves four videos comprising of two car videos, an Akiyo video, and a face video.

4.2.1.3.2 Experimental Setup

The performance of the MOTSFK object detection and the object tracking is evaluated using the experiment. Experimentation is done in the personal computer with the following specifications: windows8 operating system, 6 GB

RAM and 3.2 GHz Intel i-7 processor. The experimentation platform is the MATLAB environment.

4.2.1.3.3 Evaluation Metrics

In this section, the evaluation parameters that determine the effectiveness of visual tracking for detecting and tracking the object are presented. The evaluation metrics are the error value and the score value.

Error value

Error value is the most important parameter that determines the effectiveness of visual tracking for tracking and detecting the object. It is defined as the difference between the original image and the obtained object image. For the effective method, the error value should be less when compared with the other existing methods.

Score value

Score value is another parameter that defines the effectiveness of this method. It is defined as the similar matches existing between the original image and the targeted object image. For an effective method, the value of the score should be of the maximum value.

4.2.1.3.4 Implementation Screen Shots

This section presents the screen shots of the implementation process. Table 1 represents the screen shots of the car1 video. This screen shot is to prove the effectiveness of this method in tracking the object. The images present in frame numbers 1, 3, 5, 8, 10, 15, 20, 25, and 30 are presented in the table below and are analyzed for validating the performance of the visual tracking method and the existing method in terms of their performance metric.

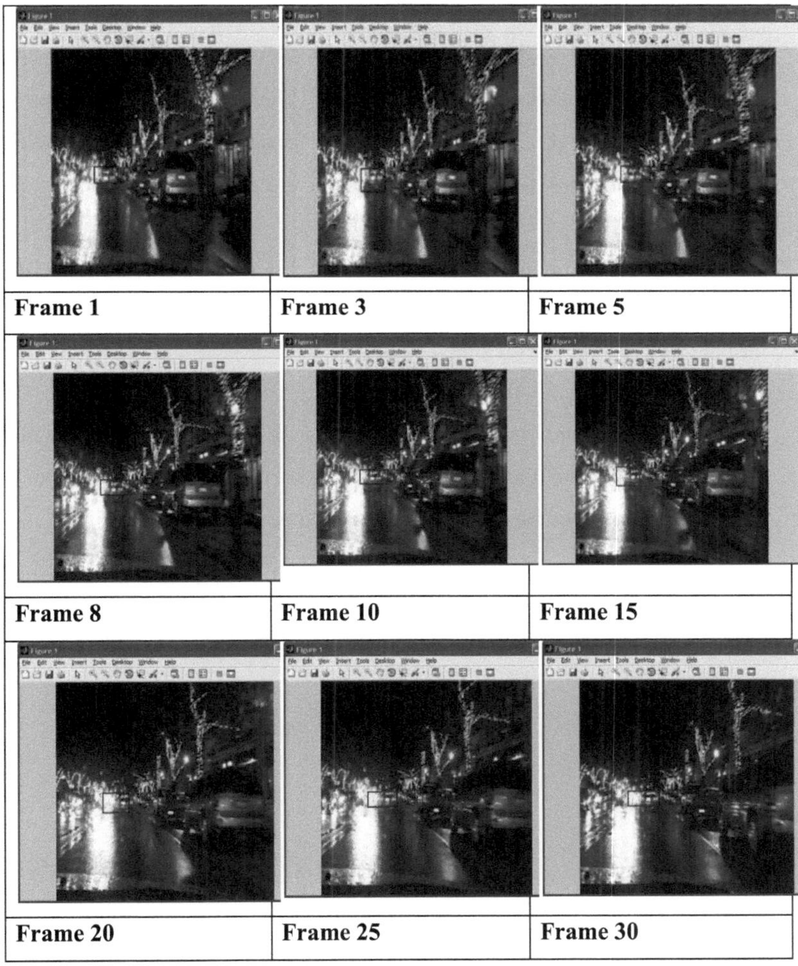

Plate 1: Implementation Screen Shots

4.2.1.3.5 Performance Analysis

In this section, the performance of Visual Tracking by Wavelet Moment, Shape and Histogram is evaluated using the performance metrics, namely, error value and score value. The performance of visual tracking is compared with Active Contour-Based Visual Tracking (Weiming *et al.*2013) to prove the effectiveness of this method. This section presents the detailed performance analysis for which four videos are used and the metrics are evaluated.

Performance analysis using car1 video: Car1 video is taken for the analysis and the performance of visual tracking is evaluated through a deep comparison of the results of the existing method, in terms of the error value and the score value.

Score value for car1 video: the score value is obtained using the visual tracking method and the existing method. For the tracking method to be effective, the value of the score obtained should be the maximum (Fig:3).

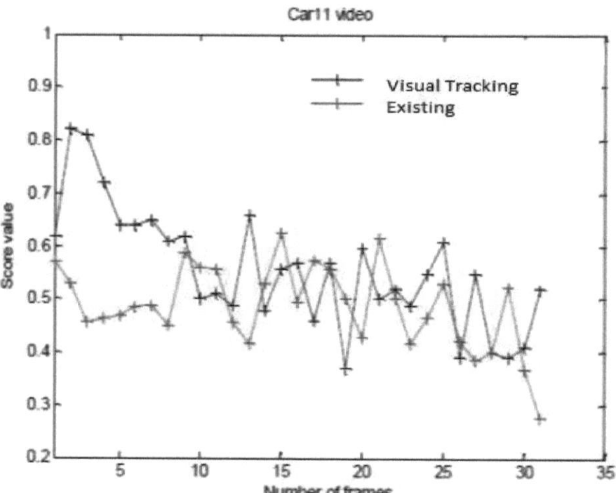

Fig 3 : Score Value For Car1 Video.

The comparison is made between the visual tracking and existing methods with the number of frames along the X-axis and the score value along the Y-axis (Fig:3). When the number of frames taken is 5, the value of the score obtained for the visual tracking method is 0.63, whereas, for the existing method, the value of the score obtained is 0.48. With 10 number of frames, the value of the score obtained for the visual tracking method is 0.5 and the value of the score obtained for the existing method is 0.57. The value of the scores obtained for the MOTSFK and the existing method are 0.57 and 0.62 respectively, when the number of frames is 15. In the graph, the value of the score value is fluctuating for both the MOTSFK and the existing methods. Similarly, for the frames 20, 25, and 30, the values of the score generated using visual tracking are 0.55, 0.62, and 0.4, and the values of the score generated using the existing method are 0.43, 0.53, and 0.28 respectively. From the above numerical interpretation, it is evident that the score value of the visual tracking method is greater when compared with the score value of the existing method, which proves the effectiveness of the visual tracking method.

Error value for car1 video: the error value of the car1 video is obtained using the visual tracking method and the existing method. The value of error obtained using the effective method should be the minimum.

The graph with the number of frames along the X-axis and the value of the error along the Y-axis is shown. The error values are fluctuating for any number of frames (Fig:4). When the number of frames is 5, the value of the error obtained using this method is 2.5 whereas, for the existing method, the value of the error is 6.5. When the number of frames is 10, the value of the error obtained using this method is 12.5 whereas, for the existing method, the value of the error obtained is 17. For the number of frames 15, the value of the error obtained for this method is 6 and the value of the error obtained for the existing method is 10. When the number of frames is 20, the value of the error obtained using the visual tracking method is 6 whereas, for the existing method, the value of the error obtained is 7. Similarly, for the number of frames 25 and 30, the value of the error obtained using the visual tracking and the existing method are 12.5, 4 and 13, 27 respectively. As can be known early an effective method is the one that possesses the minimum value of error; the above-explained numerical values denote that the value of error using visual tracking is less when compared to the value of error obtained using the existing method. It is clear that visual tracking is an effective method compared to the existing method.

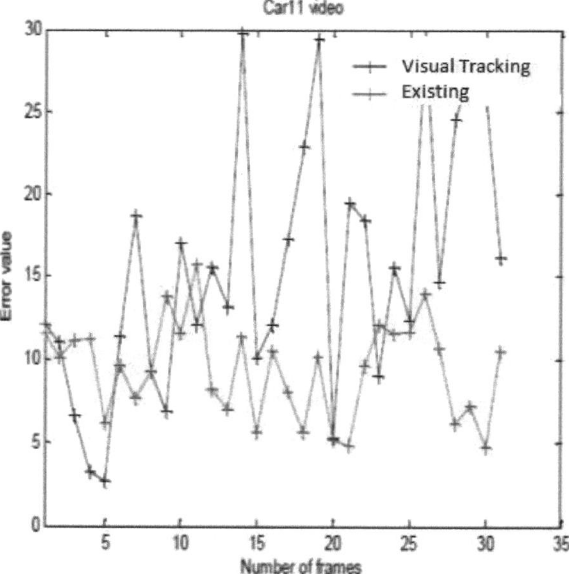

Fig 4 : Error Value For Car1 Video.

Performance analysis using car2 video: This section depicts the performance analysis of the visual tracking method and the existing method using the car2 video. The parameters considered for analysis include the error value and the score value. The effectiveness of the method is explained in the following sections.

Score value of the car2 video: the values of the score obtained using the visual tracking method and the existing method are shown in the graph. The high value of the score denotes the effectiveness of the method (Fig:5).

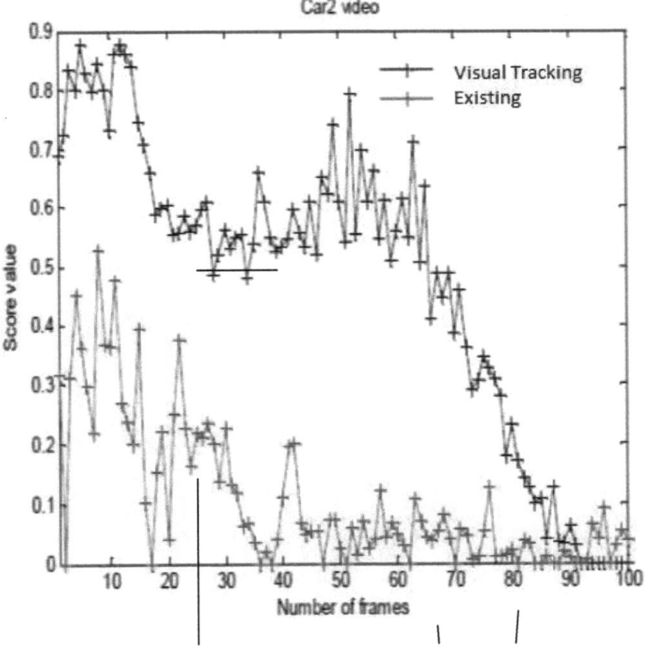

Fig 5 : Score Value For Car2 Video.

The graph with the number of frames along the X-axis and the value of the score along the Y-axis is analysed. For analysis, the number of frames is set to vary from 5 to 100. The score value of visual tracking and the existing method are compared to prove the effectiveness of the visual tracking method. The effectiveness is simply the quality of the object tracked. If the score value is greater, then the quality of the object tracked will be good. The value of the score is varying for different number of frames. When the number of frames is 10, the value of the score obtained using the visual tracking method is 0.73 whereas, for the existing method, the value of score obtained using the existing method is 0.44. The value of score obtained using the visual tracking method is greater when compared with the value of the score obtained using the existing method. Furthermore, on increasing the number of frames, following are the value of the score obtained for the visual tracking method and the existing methods. When the number of frames is 20, the values of score obtained are 0.55 and 0.23 for the MOTSFK and the existing methods respectively. For the number of frames 30, the values of the score obtained using the visual tracking method and the existing method are 0.55 and 0.22 respectively. The values of score for the visual tracking method and the existing method are 0.55 and 0.1 respectively, for the number of frames 40. For the number of frames 50, the values of score for the visual tracking method and the existing method are 0.54 and 0.01 respectively. Similarly for the other number of frames, namely 60, 70, 80, 90, and 100, the values of score obtained using the visual tracking method and the existing method are 0.54, 0.4, 0.18, 0.02, 0 and 0.02, respectively. From the above numerical interpretation, it is very clear that the visual tracking method possesses a high value of score when compared with the score value of the existing method.

Error value of the car2 video: The error value is obtained for the visual tracking method and the existing method for a number of the frames. For the effective method, the error value will remain the minimum. When the number of frames is 10, the value of the error obtained using the visual tracking method is 4 whereas, for the existing method, the value of the error obtained is 10. With 20 frames, the value of the error obtained using the visual tracking method is 10 whereas for the existing method, the value of the error obtained is 20. This value itself decides that the error value of the visual tracking method is less than that of the existing method. When the number of frames is

30, the value of the error obtained using the visual tracking method is 20 whereas for the existing method, the value of the error obtained is 40. Similarly, for 40 frames, the value of error obtained using the visual tracking method is 20 and for the existing method, the value of the error obtained is 80. The value of error plotted in the graph for the number of frames 50 is 20 for the visual tracking method and 35 for the existing method. In the same way, the value of the error obtained using the visual tracking method is 20 and using the existing method is 40 for 60 numbers of frames. When the frame number remains at 70, the values of error obtained using the visual tracking method and the existing method are 30 and 50 respectively. Similarly, for 80, 90, and 100 frames, the values of error generated using the visual tracking method are 50, 76.5, and 100 respectively and the values of error generated using the existing method are 70, 70, and 180 respectively. It is clear that the visual tracking method possesses less error value when compared with the existing method, which proves the effectiveness of the visual tracking method (Fig:6).

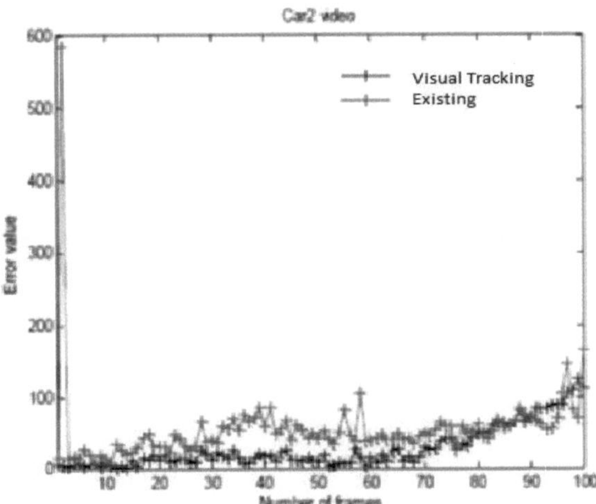

Fig 6 : Error Value For Car2 Video.

Performance analysis using face video: This section depicts the performance analysis of the visual tracking method and the existing method in terms of the performance metrics, namely, the error value and the score value using the face video. The Following shows the deep insight analysis using the face video. The performance analysis of the error value and the score value with the clear comparison of the visual tracking method with the existing method is shown.

Score value of the face video: the performance analysis of the face video uses the visual tracking method and the existing method in terms of the performance metrics score value. Score value is nothing but the quality of the image or the matching of the image when compared with the original image. The effectiveness of the method is found when the value of the score obtained is the maximum (Fig:7).

With the number of frames 10, the value of the score obtained using the visual tracking method is 0.36 and the value of the score obtained using the existing method is 0.18. The values of the score obtained using the visual tracking method and the existing method are 0.38 and 0.1 respectively when the number of frames is 20. With the number of frames 30, the visual tracking method generates a score value of 0.33 and the existing method generates a score value of 0.13. When the number of frames considered is 40, the score value obtained using the visual tracking method is 0.53 and the value of score generated using the existing method is 0.15. From the above numerical value, it is concluded that the value of the score for the existing method is less when compared with the visual tracking method. The visual tracking method finds a vast deviation in the value of score from all the existing methods for any number of frames. Let us now discussion the score value of the visual tracking method and the existing methods based on the number of frames 50, 60, 70, 80, 90, and 100. When the number of frames is 50, the value of the score obtained using the visual tracking method is 0.35 whereas the score value obtained using the existing method is 0.14. At the same time, when the score value obtained using the visual tracking method when the number of frames is 60 is 0.57 and the score value of the existing method is 0.11. When the number of frames is 70, the value of the score obtained using the visual tracking method is 0.3 and using the existing method, the value of score obtained is 0.18. When the number of frames is 80, the value of the score obtained using the visual tracking method is 0.32 and the score value obtained using the existing method is 0.15. Likewise for the number of frames 90 and 100, the value of score obtained using the visual tracking method is 0.25 and 0.55 whereas using the existing methods, the value of score obtained is 0.2 and 0.15 respectively. It is clear that the visual tracking method possess an improved score value when compared with the existing method. The higher value of score shows the effectiveness of the visual tracking method when compared with the existing method. (Please see the previous paragraph and carry out the corrections).

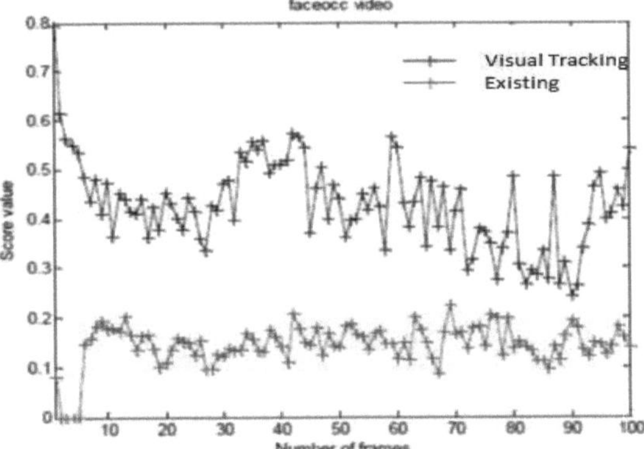

Fig 7 : Score Value For Face Video.

Error value of the face video

This section presents the error value of the face video obtained using the visual tracking method and the existing method. The Error value of the face video obtained using an effective method will be the minimum when compared with the other methods. The error value of the face video that depicts the performance analysis of the visual tracking method and the existing method with respect to the number of frames is given below (Fig:8).

The error values of the visual tracking method and the existing method are illustrated for various values of the number of frames. When the number of frames is 10, the value of error obtained using the visual tracking method is 7 whereas the error value obtained using the existing method is 10. The value of the error with respect to the number of frames 10 shows that the visual tracking method possesses minimum value of the error when compared with the existing method. Furthermore, when the number of frames is 20, the value of error obtained using the visual tracking method is 25, and the value of the error obtained using the existing method is 45. For the number of frames 30, the value of error obtained using the visual tracking method is 40 and the value of the error obtained using the existing method is 70. For the number of frames 40, the value of the error obtained using the visual tracking method is 20 and the value of the error obtained using the existing method is 60. When the number of frames is 50, the value of the error obtained using the visual tracking method is 60 and the value of the error obtained using the existing method is 70. For the number of frames 60, the value of the error obtained using the visual tracking method is 30 and the value of the error obtained using the visual tracking method is 50. When the number of frames is 70, the value of the error obtained using the visual tracking method is 47 and the value of the error obtained using the existing method is 60. For the number of frames 80, the value of the error obtained using the visual tracking method is 10 and the value of the error obtained using the existing method is 12. When the number of frames is 90, the value of the error obtained using the visual tracking method is 30 and the value of the error obtained using the existing method is 45. When the number of frames is 100, the value of the error obtained using the visual tracking method is 20 and the value of the error obtained using the existing method is 35. From the above analysis, it is very clear that the value of the visual tracking method takes a low value of the error and the existing method possesses a higher value. The low value of the error indicates the effectiveness of the visual tracking method. (See the previous paragraphs).

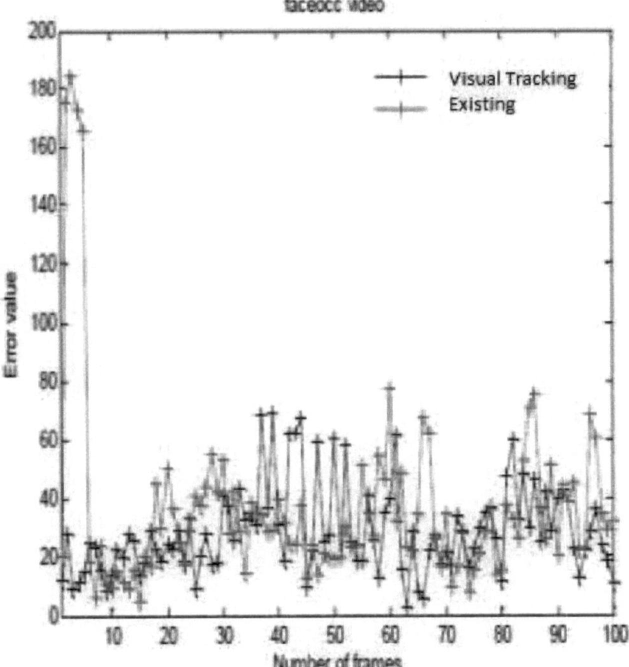

Fig 8 : Error Value For Face Video.

Performance analysis using akiyo video

This section depicts the performance analysis of the visual tracking method and the existing method, using the akiyo video. The score value and the error value obtained using the visual tracking method are compared with that of the existing method. The low value of the error and the greater value of the score decide the effectiveness of the visual tracking method.

Score value of the akiyo video

The score value of the akiyo video and the performance analysis is described as follows. When the number of frames is 10, the value of the score obtained using the visual tracking method is 0.8, and the value of the score obtained using the existing method is 0.35. When the number of frames is 20, the value of the score obtained using the visual tracking method is 0.78 and the value of the score obtained using the existing method is 0.38. The value of the score obtained using the visual tracking method is 0.65 and the value of the score obtained using the existing method is 0.55 when the number of frames is 30. Similarly for the number of frames is 40, the value of the score obtained using the visual tracking method is 0.9 and the value of the error obtained using the existing method is 0.49. For the number of frames 50, the value of the error obtained using the visual tracking method is 0.58 and the value of the error obtained using the existing method is 0.5. Likewise for the number of frames 60, 70, 80, 90, and 100, the value of the score obtained using the visual tracking method is 0.7, 0.72, 0.78, 0.8, and 0.7 whereas, the value of score obtained using the existing method is 0.55, 0.55, 0.4, 0.55, and 0.3 respectively. It is clear that the value of score obtained using the visual tracking method is greater than the score value of the existing method, which proves the effectiveness of the visual tracking method(Fig:9).Correct as above)

Error value of the akiyo video

The error values of the visual tracking method and the existing method using the akiyo video are compared. The error value is analyzed with respect to the number of frames. When the number of frames is 10, the value of the error obtained using the visual tracking method is 20 whereas the error value obtained using the existing methods is 40. Similarly, for the number of frames 30, 50, 80, and 100, the value of the error obtained using the visual tracking method is 20, 13, 4, and 19 respectively whereas the value of the error obtained using the existing method is 35, 35, 17, and 38 respectively. The performance analysis with the parameter error value indicates that the value of the error for the visual tracking method is less when compared with the error value of the existing method. It is clear that the visual tracking method is effective over the existing method (Fig: 10). (Same as the previous paragraphs)

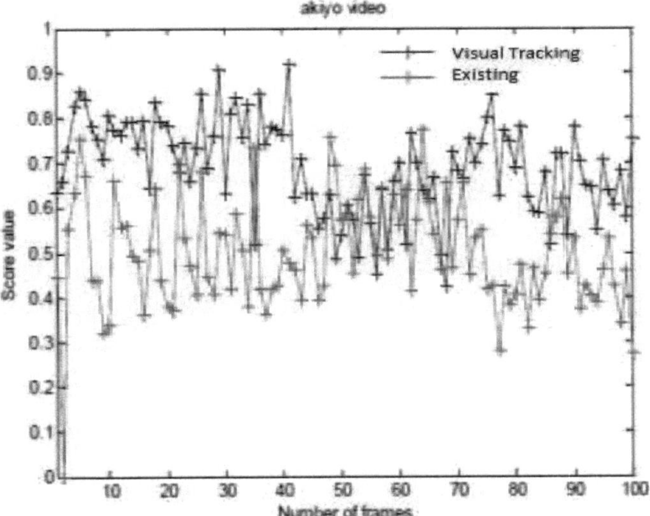

Fig 9 : Score Value For Akiyo Video.

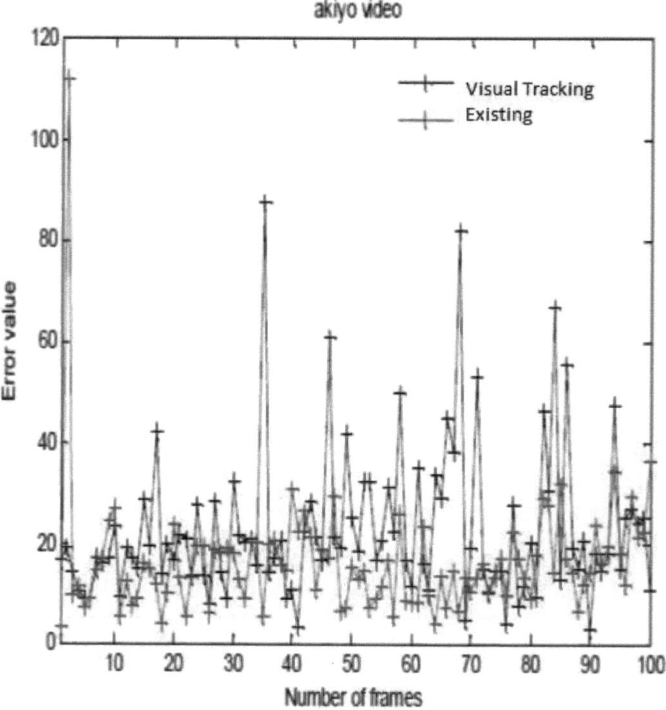

Fig 10 : Error Value For Akiyo Video.

The effectiveness of the visual tracking method is proved in this section through table 5.2. For a method to be effective the score value should be the maximum and the error value should be the minimum. Effectiveness, in other words, is described as the quality of the object tracked using the visual tracking method. For proving the effectiveness, four videos are considered, namely, car1 video, car2 video, face video, and the Akiyo video. The numerical comparison is as follows. For car1 video, the values of score and of error obtained using the visual tracking method are 0.83 and 2.5 respectively. But for the existing method, the value of the score and the error obtained are

0.62 and 4 respectively. For the car2 video, the value of the score and error obtained using the visual tracking method is 0.88 and 4 whereas, using the existing method, the value of score and error obtained is 0.54 and 10 respectively. Similarly, for the face video, the value of the score and the value of the error obtained using the visual tracking method is 0.8 and 12 respectively. But for the existing method, the value of the error and score is 0.24 and 5 respectively. Likewise for the Akiyo video, the value of score and error obtained using the visual tracking method and the existing methods are 0.93, 4 and 0.78, 5 respectively. From the above comparison, it is very clear that the value of score obtained using the visual tracking method is greater than the value of score obtained using the existing method. The score of the visual tracking method is greater for all the videos. At the same time, the error value of the visual tracking method obtained for all the videos is less than that of the error value of the all videos obtained using the existing method, which proves the effectiveness of the visual tracking method. In other words, the quality of the object tracked using the visual tracking method remains high when compared with the quality of the object obtained using the existing method.(correct as the previous para)

Table1: Comparison Table With The Score Values And Error Values of The Visual Tracking And The Existing Methods.

Videos	Visual Tracking Method		Existing Method	
	Error value	Score value	Error value	Score value
Car1 video	2.5	0.83	4	0.62
Car2 video	4	0.88	10	0.54
Face video	12	0.8	5	0.24
Akiyo video	4	0.93	5	0.78

4.2.2 Multiple Object Tracking By Employing Shaped Based Features And Kalman Filter (MOTSFK)

Moving object tracking is the current research area that attracts the scientists in contributing a lot towards the rising challenges existing in the field of tracking   moving objects. The major challenge faced in object detection from   static images is that, it requires a specific object detection system to differentiate the particular group of objects from all other different groups. Moreover, differentiating the objects under varying conditions of illumination, rotation, and occlusion is the major task that requires a robust object detection system. The presence of noise in the tracking environment creates problems with detection. There are a large number of algorithms to solve these challenges persisting in object tracking along with their strengths and weaknesses. In the existing works, the methods used for object detection and tracking of the objects are background based object segmentation and feature extraction. The major drawback of the above-mentioned methods is that they tracked only a single object, which is not suitable for videos with a number of objects. Therefore, to improve the performance of object detection and tracking, a highly robust method named as the multiple objects detection and tracking MOTSFK, is proposed. The importance of this method is that it uses shape-based features and the Kalman filter.

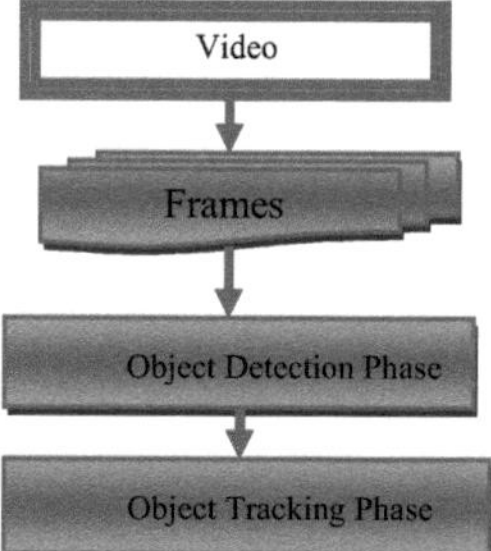

Fig 11 : Block Diagram of The MOTSFK Technique

The block diagram shows the various steps involved in the MOTSFK multiple objects detection and tracking method. Initially, this technique converts the videos into frames and then detects the object in the videos. The main advantage of this method is that it is capable of detecting multiple images in the video which is the major challenge of the existing methods. Then the detected objects are tracked using the MOTSFK technique (Fig : 11)

4.2.2.1 Converting Video to Frames

Conversion of the input videos into frames is the first step involved in the MOTSFK multiple objects detection and tracking. For performing this conversion, the conversion step uses a technique named as shot segmentation , which is the most commonly employed method for converting the videos into frames and it is used in most of the applications like video retrieval, video watermarking and video summarization video retrieval. In this shot segmentation technique, the first step is that the video is converted into a number of non-overlapping units. Each overlapping unit is termed as the shot. One can find a remarkable modification in every shot and it is known that the modification of a shot from others is contributed by other objects in the video and due to background noise. A number of video shot segmentation techniques are available in the literature. In this chapter, the shot segmentation method used is the wavelet transform-based shot segmentation. Distance measure and transformation are the two parameters that determine the number of frames present in the individual shot. In this method, initially, the first two frames are separated and they are presented as a set of blocks. For each block of the frame, DWT is applied. Wavelets are nothing but the mathematical functions whose average value is zero in a defined restricted interval. Let us consider the input that possesses a sequence of m real numbers which can be denoted as

$y_1, \ldots\ldots, y_m$. Now, the DWT of the n sequence of the x real numbers is given in the following formula as,

$$w_{nx} = \int_{-\infty}^{+\infty} y(t)\psi_{nx}(t)\,dt \qquad (4.17)$$

where, $\qquad \psi_{nx}(t) = \dfrac{1}{\sqrt{\alpha_{jk}}}\,\psi\left[\dfrac{t-\beta_{nx}}{a_{nx}}\right] \qquad (4.18)$

ψ is the mother wavelet, α is the scaling parameter, and β is the shifting parameter. Initially, the distance between the first two frames is calculated. The distance measure is represented as d_{ij}, which is the distance between the frames F_i and F_j respectively. The distance between the frames is given by,

$$d_{ij} = \sqrt{\sum_{i=1}^{m}(F_i - F_j)^2} \qquad (4.19)$$

Initially, the distance between the first two frames is calculated using the above formula and the distance for all the other consecutive frames is also calculated using (4.3). Once the distance of the consecutive pair of frames is computed, the frames within the shots are recognized, based on the minimum distance between the frames. Minimum distance implies that there is maximum similarity existing between the frames. Let us consider the input video as represented as $V[j,k]$ and the segmented non-overlapping shots are represented as $Z(j,k)$.

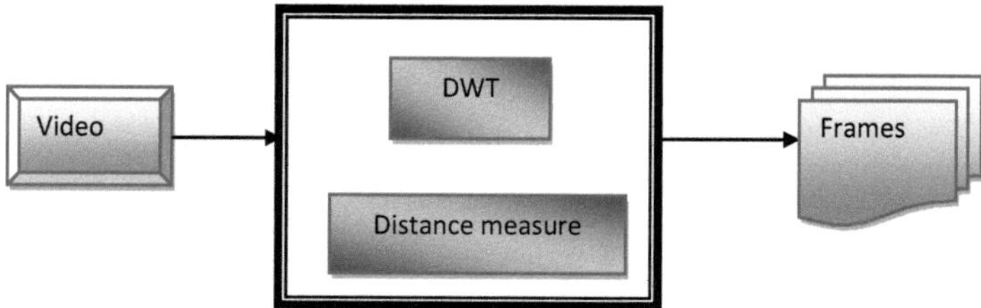

Fig 12 : Process of Video Conversion Into Frames.

Therefore, the process of video conversion involves the above steps .in figure ...Initially, the video is segmented into shots. A pair of frames is separated and is presented as blocks. DWT is applied to each block and the distance between the pair of the frames is calculated. Based on this distance measure, the minimum distance frame is recognized that has the maximum similarity, as is required (Fig12)

4.2.2.2 Object Detection Phase

This section presents the detection of the objects and the determination of the dimensions of the object. The object detection is carried out using the process, named as the manual marking process and the dimensions of the object considered are the object center, width and height (Fig:13).

At first, consider the first frame in the shot and it is marked manually. To identify the significant area, a cross check is allowed by introducing an area constraint. The main condition of the area constraint is to allow checking whether the area of the object in the frame is the minimum and lies within the restricted area. The restricted area ensures that the noise in the frame is reduced. The notation for the minimum allowable area is Th_{area}, the area of the object pixel is O_b and the area of the object in the frame is $Area_i$. Now, the condition of the area constraint can be stated as the following,

$$if\ Area_i < Th_{area,} \qquad then\ discard\ O_b \qquad (4.20)$$

Equation 4.20 explains that the clusters without the minimum area will not be taken as objects and therefore, the objects less than the minimum area Th_{area} are discarded. Thus, all the objects with the area within the permissible range are selected and the dimension is determined. Initially, the corners of the object are detected and the center of the object is determined. Let us represent the detected object as O_b, which is the set of all the detected objects from the video. The detected objects are represented as given below,

$$O_b = \{O_b^1, O_b^2, O_b^3, ..., O_b^m\} \qquad (4.21)$$

Initially, the dimension of the object is fixed, namely the corners of the object. Therefore, the corners of the object are given as $a_{x,y}, b_{x,y}, c_{x,y}, and\ d_{x,y}$. The corners are necessary for determining the center point of the object and the center point of the object is represented as $cen_{x,y}$. The center point $cen_{x,y}$ is formulated as,

$$cen_x = \frac{(a_x + b_x + c_x)}{4} \text{ and } cen_y = \frac{(a_y + b_y + c_y)}{4} \qquad (4.22)$$

Once the center points are fixed using the corners, it is necessary to determine the height and width of the object. The height and width calculation depend on the edges and it is determined, using the difference between the top and bottom edge, whereas the width is calculated using the difference of the right and   left edges of the object. Moreover, the tilt of the object is also determined. Let us consider the right and left edges of the image as X_R and X_L respectively. Similarly, the top and bottom edges are represented as X_T and X_B respectively. Now, the width of the object image is computed as,

$$Width = |X_R - X_L| \qquad (4.23)$$

The height of the object is denoted as shown below,

$$Height = |X_T - X_B| \qquad (4.24)$$

The detected object point is represented as O_b^i and each point in the detected objects is represented as $O_b^i = \{cen_{x,y}^{(i)}, width_i, height_i\}$.

where, i is the number of the object and it ranges from $1 \leq i \leq M$, O_b^i is the ith object in the detected set of objects. $width_i$ is the width of the ith object and $height_i$ is the height of the ith object.

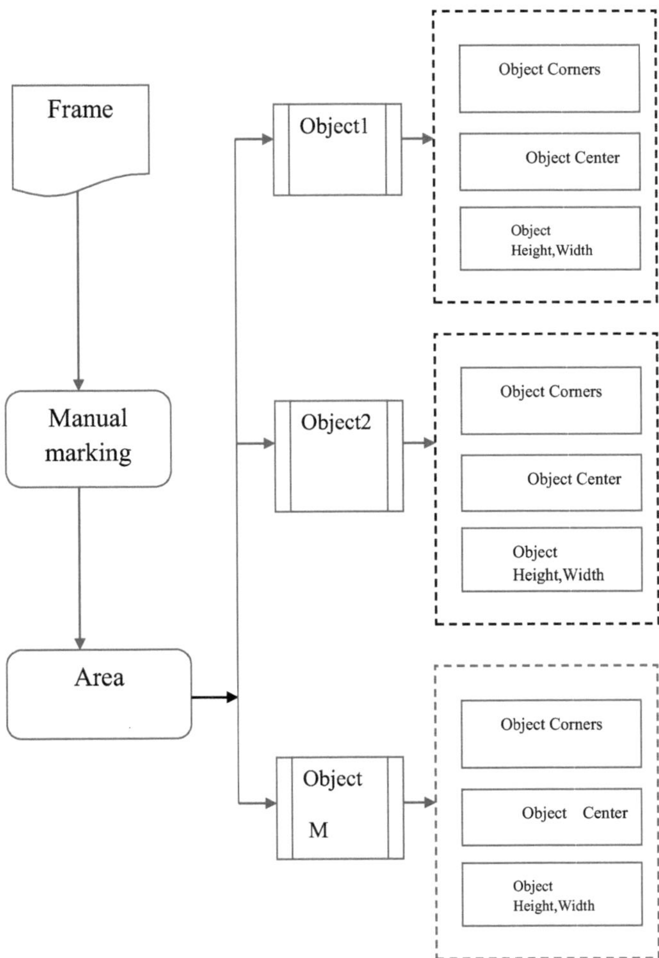

Fig 13 : Object Detection Phase

The above figure explains the overall operation of the detection phase. Initially, the frames in the shots are marked manually and the area constraint is introduced. This area constraint is used to detect the object with the minimum area that is found as the significant object. Once, the object is determined, the corners are determined so as to calculate the object centre and tilt. In addition, the height and width of the object are determined using the edges. Thus, all the objects in the frame are determined and enter into the tracking phase.

4.2.2.3 Object Tracking Phase

In this section, object tracking is discussed in detail. Once the object in the frame is located, the detected object in the frame is tracked by locating the position of the object in the successive frames. First, the first frame of the video is extracted and the object is marked manually, and the position of the object in the frame is determined. The object detection is discussed in the previous sections. These detected objects in the frames are followed in the other frames to keep track of the object. Similarly, this tracking principle is carried out for all the other objects in the frames. The flow chart of the tracking phase explains the tracking principle clearly (Fig:14).

Initially, the process begins with the single object and the Kalman filter is applied to it for predicting the new centroid of the object. The concept is that it selects the object based on the centroid of the object. If the centroid of the object computed using the Kalman filter and the old centroid are found to be the same, then keep the location of the object the same as the current location. But when the centroid values of the object vary, then, recompute the location of the object as shown in the flowchart.

Kalman filter: Kalman filtering is the efficient method of measurement that uses a series of measurements found at a particular time, and contributes towards determining the unknown variables. The estimation of the unknown variables using the Kalman filter is very precise even under the midst of noise and other inaccuracies and it is highly advantageous over the other methods, in terms of precision. The precision is high for the estimation of unknown variables using a series of measurements rather than using a single measurement. Moreover, the Kalman filter is the recursive estimator. In other words, the recursive estimator or the Kalman filter uses the estimated state from the previous iteration and the measurement of the current iteration, to estimate the current state. Kalman filtering involves two steps.

The basic principle of the operation of the Kalman filter is described below. The first step is the prediction step and in this step, the filter estimates the present state variable. Based on the measurement of the next iteration, the estimated state variable from the previous iteration is updated. Updating the estimated state variable of the previous iteration depends on the weighted average that indicates that the weight of the estimated state variable is increased with high certainty. In this work, the Kalman filter estimates the new centroid for an object in the frame.

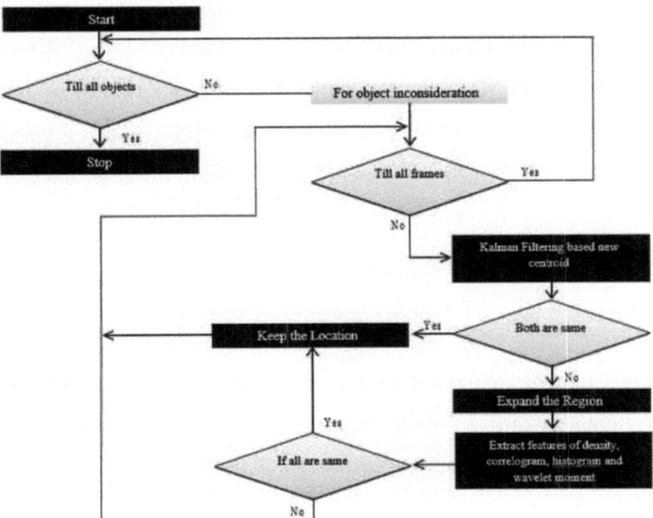

Fig 14 : Flow Chart of The Object Tracking Phase.

Once the Kalman filter predicts the new centroid, the old centroid of the object and the new centroid of the object are compared. If both the centroid values of the object are the same, then locate the position of the object the same as the previous position. If the centroids are not the same, then determine the new dimension. The dimension of the image is extended at all its edges. The edges of the image at the initial position are noted as $X_R(i), X_L(i), X_B(i)$, and $X_T(i)$, and the extended edges are denoted as $U_R(i), U_L(i), U_B(i)$, and $U_T(i)$.The notations $X_R(i), X_L(i), X_B(i)$, and $X_T(i)$are the edges on the right, left, bottom, and top of the image in the previous state whereas, $U_R(i), U_L(i), U_B(i)$, and $U_T(i)$denote the extended edges on the right, left, bottom, and top of the image. With this extended dimension, a new image is formed. Features are extracted from this newly cropped image and they include the density, correlogram, and histogram and wavelet moment. The features that are to be extracted from the newly cropped image are discussed in detail below.

Density: Density is denoted as the number of pixels per inch (PPI) or pixels per centimeter (PPCM). From the name, it is clear that density is defined as the number of pixels per unit measure present in the cropped image. Let us consider the number of pixels present in the cropped image g, the centimeter covered is h, and the PPCM density is PD. Now, the PPCM density is formulated as,

$$PD = \frac{g}{h} \tag{4.25}$$

Correlogram: Correlogram represents the autocorrelation plot of the cropped image. In other words, a correlogram is defined as the cross-correlation of the image itself. The representation of the autocorrelation of the image $f(x,y)$is represented as,

$$f(x,y) * f(x,y) = \int_{-\infty}^{\infty} \int_{-\infty}^{\infty} f(x',y')\, f_i(x+x',y+y')\, dx'\, dy' \tag{4.26}$$

where, $f(x,y)$ is the image that denotes the two-dimensional brightness and $f(x',y')$ is the dummy variables in the integration.

Histogram: Histogram is the third feature that is extracted from the image. The distribution of the pixels in the image is represented graphically using the histogram, and it represents all the tabulated frequencies that are graphed over the discrete interval. The area of the tabulated frequencies that are distributed over the discrete intervals is directly proportional to the frequency of the pixel in the cropped image.

The most important feature extraction method employed for the extraction of the features from the newly cropped image is the Image Moments method, which is capable of generating highly discriminative features. The wavelet moment is the feature extraction method that belongs to the orthogonal moment family and it employs the kernel function, named as the orthogonal wavelet function. The wavelet moment is advantageous as it uses the merits of the wavelet and moment analysis for constructing the improved version of the moment descriptors. The wavelet moment for the image $f(x,y)$ of dimension $(m \times m)$ is represented as,

$$M = \sum_{x=1}^{m}\sum_{y=1}^{m} \psi_{ab}(r)e^{-jq\varpi} f(r,\varpi) \tag{4.27}$$

where, $r = \sqrt{x^2 + y^2}$, $\varpi = \tan arc\left(\dfrac{y}{x}\right)$, and $\psi_{ab}(r)$ are the basics of the mother wavelet. Therefore, features from the expanded cropped image are extracted and the extracted features include the density, Correlogram, histogram and wavelet moments, and these features are represented as $PD_i, AC_i, H_i, and M_i$. These features are determined for various locations of the expanded cropped image. From all the features that are determined for the various locations, only the features that take the high value of density, histogram, and the wavelet moment and the low values of the correlogram features are selected. If the above conditions are satisfied then the location of the cropped image is kept the same and the process is continued for the next frame. The condition is represented as,

$$if\ \lambda(\max PD_i) = \begin{pmatrix} \lambda\left(\min AC_i\right) = \lambda(\max H_i) = \lambda(\max M_i), \\ Then\ take\ the\ location \end{pmatrix} \qquad (4.28)$$

where, λ is the location, max denotes the maximum of all the values, and min is the minimum of all the values.

The above process is carried out for all the other frames and in the following steps, the second frame is taken as the reference and the third frame is cropped from the image based on the reference image. In general, the reference image is the ith image and the second image that is cropped depends on the $i+1^{th}$ image and the cropped image is saved. The location of the cropped image is tracked and the same process mentioned above is repeated for all the other detected objects.

4.2.2.4 Experimental Results

This section presents the results and discussion of Visual Tracking by the Wavelet Moment, Shape and Histogram of tracking multiple objects in the video. Figures 4.5 and 4.6 denote the input image with the sample input video images and the tracked original images. Initially, the sample images are provided for tracking the objects and these images are used as the reference images in tracking the objects. The frames are analyzed for locating the target object for which the reference image is used. The perfect match of the object in each of the frames is marked and the location of the object in the video is located.

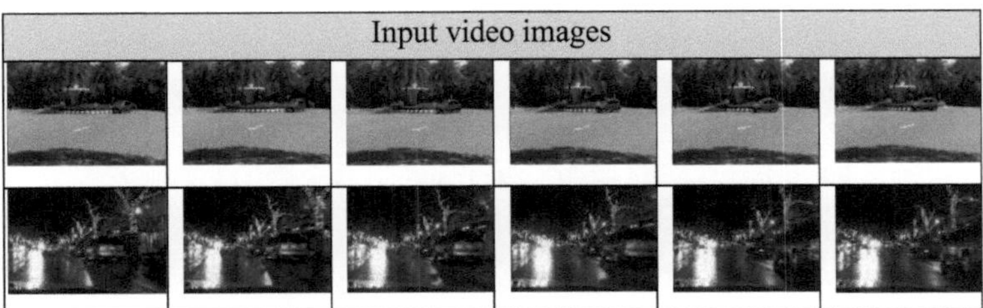

Plate 2 : Sample Input Video Images

Plate 3 : Original Image With Tracking Result

4.2.2.4.1 Performance Analysis

This section presents the performance analysis of the Multiple Object Tracking by Employing Shaped Based Features and the Kalman Filter (MOTSFK) in terms of their efficient and accurate tracking performance. The performance analysis of the MOTSFK in terms of the error value and score value describes the effectiveness of this method. The performance results of MOTSFK are compared with those of the existing method to verify the effectiveness of this method in tracking multiple objects from the video frames.

Performance analysis using the car1 video: The performance analysis using MOTSFK and the existing method in terms of the error value and the score value is presented in the following explanations.

Error value for the car1 video: the error response of the car1 video object tracking using MOTSFK and the existing method.

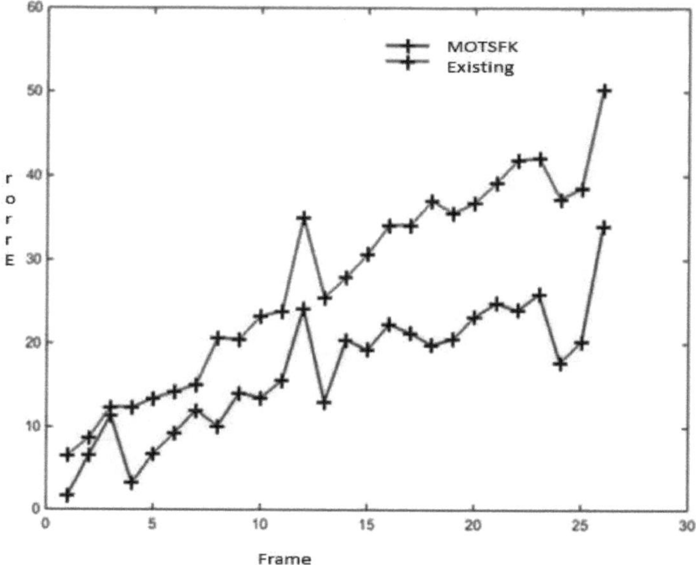

Fig 15 : Error Value of The Car1 Video

The variation of the error with respect to the number of frames is depicted in the Fig:15 When the number of frames is 1, the error value obtained with object tracking using this tracking method is 1 whereas, the error value for the object tracking using the existing method is 7. Similarly, the error value of tracking obtained using the visual tracking method when the number of frames is 5 is 10 but the existing method attains 13. The error value obtained using the visual tracking method is 15 and for the existing method, the error value is 20 when the number of frames is 10. Similarly, for the number of frames 15, 20, and 25, the error value obtained using the visual tracking method and the existing methods are 22, 20, 30 and 25, 27, 50 respectively. From the above-mentioned numerical values, it is evident that the visual tracking method outperforms the existing method. The error associated with multiple object tracking is low for the visual tracking method when compared with the existing method. Therefore, the visual tracking method possesses the capacity to track multiple objects from the video frames with good accuracy (Fig:17).

Error value for the car2 video

The error value of the car2 video using the visual tracking method and the existing method with respect to the number of frames is given below. The error value is small for the effective method. When the number of frames is 1, the error value for the visual tracking method is 4.5 and for the existing method, the error value is 4.9. The error value for the visual tracking method and the existing method when the number of frames is 5 is 2.5 and 7 respectively. Similarly, the visual tracking method attained an error value of 7.5 for the number of frames 10 and the existing method attained the error value of 7.9 for the same number of frames. The error values obtained for the visual tracking method are 3, 10, 13 and 20 for the number of frames 15, 20, 15, and 30 respectively. The existing method attained error values of 10, 13.5, 15, and 45 for the number of frames 15, 20, 15, and 30 respectively. From the numerical comparison, it is very clear that the visual tracking method attained a better performance value when compared to the existing methods. The error value is less for the visual tracking method when compared with the existing method (Fig:18).

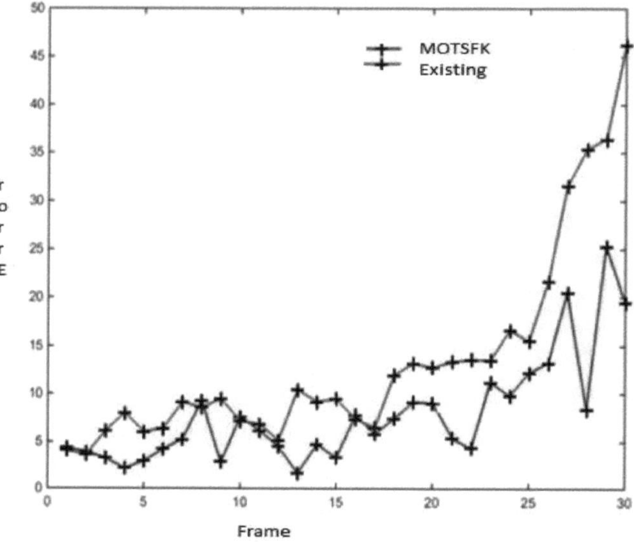

Fig16.Error value of car2 video

Thus, it can be concluded that the performance analysis of the visual tracking method is effective that it yielded a minimum error percentage when compared with the existing method. The minimum error percentage of the visual tracking method is 33% whereas the error percentage of the existing method is 50% for the multiple object tracking obtained using the car1 video. For the car2 video, the minimum value of error attained using the MOTSFK tracking method is 25% whereas for the visual tracking method, the error percentage is 48%. On these comparisons, it can be said that the visual tracking method is an effective method in multiple object tracking with the minimum error percentage.

Performance analysis in terms of score value

This section presents the performance analysis of the visual tracking method and the existing methods in terms of the score value for which, the car2 video is presented.

Score value for the car1 video

The score values of the car2 video using the visual tracking method and the existing method are compared. For an effective method, the score value will remain high. In other words, the effective means of multiple tracking remains high for the effective method. The score value as a result of multiple object tracking of the car1 video with respect to the number of frames is as follows: When the number of frames is 1, the score value of multiple object tracking is 0.97 for the visual tracking method and for the existing method, the score value is 0.01. When the number of frames is 5, the score value obtained using the existing method is 0.95 whereas for the visual tracking method it is 0.04. When the number of frames is 10, the score value obtained using the visual tracking method is 0.8 whereas for the existing method, the score value is 0.09. For the number of frames 15, 20, and 25, the values of the score obtained using the visual tracking method are 0.75, 0.7 and 0.6 respectively. The score value obtained for the existing methods for the number of frames 15, 20, and 25 is 0.1. From the graph, it is well understood that the score value of the visual tracking method is high when compared with that of the existing method. The high score denotes that the visual tracking method is highly efficient in tracking multiple objects from the video when compared to the existing method (Fig:17).

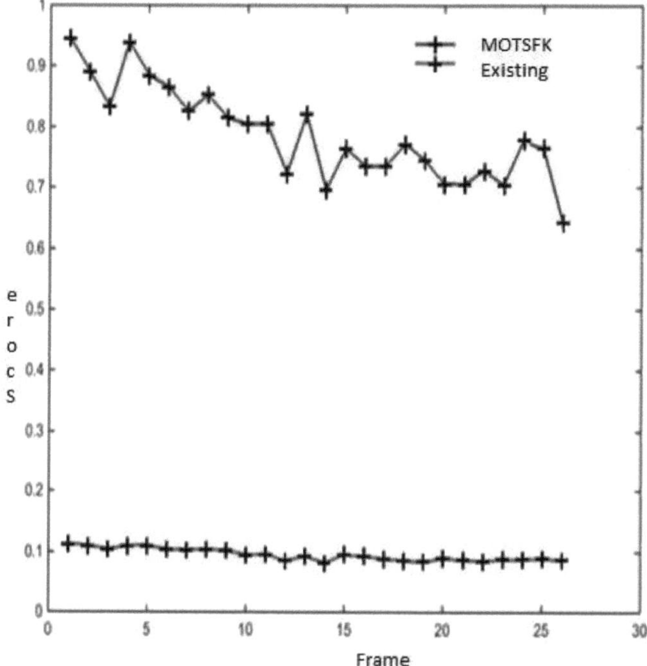

Fig17: Score value of the car1 video

Score value for the car2 video

The score value of the car2 video using the visual tracking method is compared with that of the existing method. The performance analysis defines the efficiency of the visual tracking method. For the number of frames 1, the score value obtained using the visual tracking method is 0.88 whereas for the existing method, the score value obtained is 0.85. The score values obtained using the MOTSFK multiple object tracking method and the existing method are 0.95 and 0.75 respectively, when the number of frames is 5 (Fig: 18).

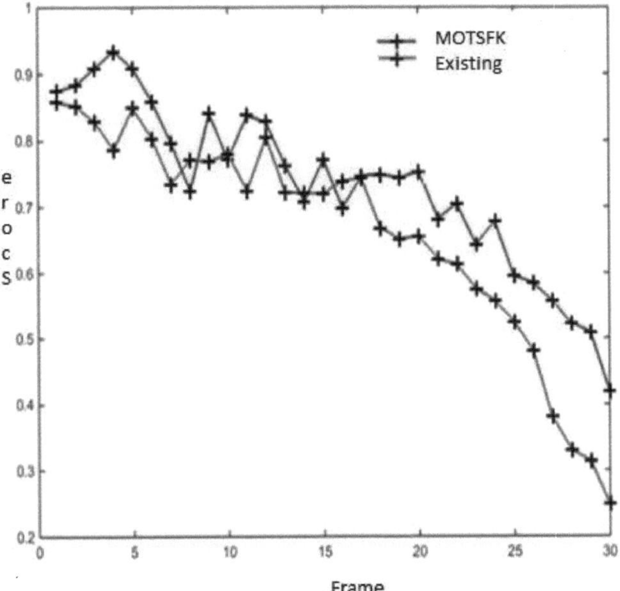

Fig 18: Score value of car2 video

For 10 frames, the score value of the multiple object tracking obtained using the visual tracking method is 0.79 and the score value attained using the existing method is 0.75. Similarly, the score values obtained using the visual tracking method and the existing method for the number of frames 15, 20, and 25 are 0.8, 0.7, 0.5 and 0.73, 0.65, 0.45 respectively. For the number of frames 30, the score value obtained using the MOTSFK multiple tracking method is 0.45 and score value using the existing method is 0.25.. It is very clear that the visual tracking method attained a good score value compared with the existing method. Initially, the score value fluctuates. The clear idea is that for less number of frames, the score value is high and it reduces with the increase in the number of frames. Moreover, the visual tracking method provided a better tracking performance compared with the existing method. (If there is only one other method, simply say 'the existing method'. If you say 'other' it means 'others'; that is 'many'. Be clear in what you write.)

The performance analysis shows that the visual tracking method is more effective. The car1 video analyzed using the visual tracking method shows that the visual tracking method attained a score value of 95% while that for the existing method is 15%, which is very low. The performance analysis of the car2 video shows that the visual tracking method achieves the score percentage of 95% and the existing method attained a score percentage of 80.5%. The effectiveness of the visual tracking method is confirmed as the score value obtained using the visual tracking method is the maximum.

The visual tracking method values are compared with those of the existing method and the discussion over the topic is shown below:

Table 2: Comparison Table - Values of Error and Score of The Visual Tracking Method and The Existing Method.

Videos	MOTSFK Method		Existing Method	
	Error value	Score value	Error value	Score value
Car1 video	2	0.95	7	0.13
Car2 video	1.8	0.93	4.8	0.87

The performance analysis of the MOTSFK multiple object tracking technique   is compared with the other existing methods and provides a discussion over the effectiveness of the visual tracking method. For the performance analysis, the car1 video and the car2 video are considered. The performance metrics considered in the analysis include the error value and the score value. For an effective system, the value of error obtained will be less and the value of score obtained will be the maximum. The values of error obtained for the visual tracking method using the car1 and car2 videos are 2 and 1.8 respectively, whereas, the values of error obtained using the existing method for the car1 and the car2 videos are 7 and 4.8 respectively. The discussion table proves that the visual tracking method attains a minimum error value when compared with the existing method. It is clearly noted that with the increase in the number of frames, the value of the error is increasing but the error of the visual tracking method is less than that of the existing method. In addition, the score value obtained using the visual tracking method is greater than that of the existing method. The score value of the visual tracking method is 0.95 for the car1 video and 0.93 for the car2 video whereas, the score value of the existing method is 0.13 for the car1 video and 0.87 for the car2 video. Thus, the value of score for the visual tracking method is greater when compared with the visual tracking method. It can be seen from the above table 4.1 that the visual tracking method attained a minimum error value and a maximum score value, which proves the effectiveness of the visual tracking method.

4.2.3 Hybrid Tracking Model For Multiple Object Videos Using the Second Derivative Based Visibility Model and Tangential Weighted Spatial Tracking Model

The MOTSFK hybrid tracking model tracks the multiple target objects from the video through the visibility and the spatial models. The video with multiple objects has many frames. But each frame of the video may or may not have the target object. Since, it is necessary to detect the presence of the target object in the frame before tracking, the frame from the video is given as input to the K- means clustering algorithm. The k-means clustering algorithm segments the objects in the frame to find the object area. The object area contains the texture pixels and the neighbour pixels. After the object segmentation process, the frames are sent to the MOTSFK hybrid tracking model. The hybrid tracking model performs the object tracking through the visibility and the spatial models. The MOTSFK visibility model uses the neighbourhood search algorithm and the second derivative matched formula for the visual tracking of the target objects. The spatial tracking model uses the tangential weighted function. Finally, the MOTSFK hybrid tracking model effectively combines the results from the visibility tracking model and the spatial tracking model. The tracked output video frame has lines indicating the movement of the target object in each frame of the video (Fig:19).

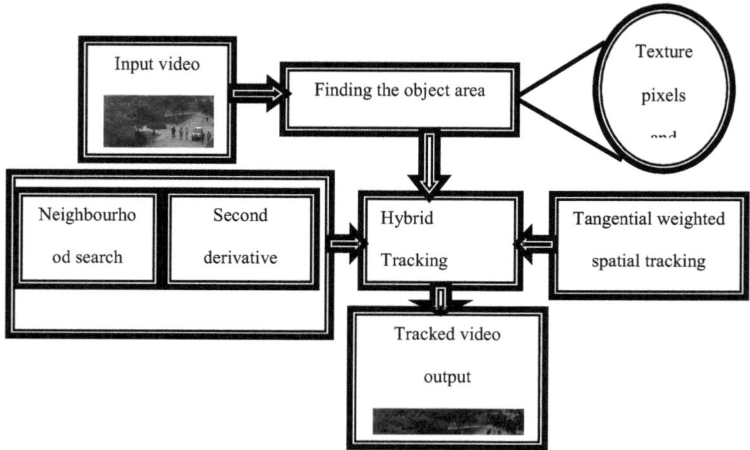

Fig 19: Block Diagram of the Visual Tracking Method

The MOTSFK hybrid tracking model performs the object tracking in the following steps,

i) The primary step is to read the input video (http://www.svcl.ucsd.edu/projects/anomaly/dataset.htm). The video contains many numbers of frames.

ii) Each frame in the video may not have the required target object. The next step is to segment the objects in the video frame. The k-means clustering algorithm extracts the object area of the video from two dimensional matrices. The object area contains the texture pixels and neighbour pixels.

iii) The segmented objects are given to the visibility tracking model. The extraction of the target objects is done based on the second derivative based visual tracking.

iv) Apply the segmented frames to the tangential weighted tracking model for the extraction of the movement of the objects within the frame.

v) In this step, combine the results of the MOTSFK visibility model and spatial tracking model.

vi) Finally, the tracked target object from the frame is obtained.

4.2.3.1 Finding the Objects from the Key Frame Using K Means Clustering

The input video has n number of video frames. The target object's presence in each frame of the video is detected through the segmentation of the objects in the frame. For this purpose, the key frames are extracted from the input video. The Next step is the detection of the presence of objects in the key frame. The k-means clustering algorithm finds the required object area to detect the objects from the key frames of the video. The K means clustering algorithm (Kanungo *et al.,* 2002, Zhong *et al.,* 2005) solves the clustering problem in the video in an efficient way. In the tracking of the multi object videos the object clustering is one of the major challenges to be considered. The k-means clustering algorithm efficiently overcomes the difficulties arsing due to the object clustering, and hence, the k means clustering algorithm is preferred in this research work. Feature extraction from the video allows the segmentation to be easier. In this research the Before Local Binary Pattern (LBP) model 1 (Zhenhua *et al.,*2010) efficiently extracts the required features for object extraction.

Feature construction

For constructing the frame feature, read the input video V and extract the frame V_i, which is defined as the i-th frame of the video, shown as follows.

$$V = \{V_i \quad 1 \le i \le n\} \qquad (4.28)$$

For Each and every frame, the pixel values can be represented as shown below.

$$V = \left\{ v_i^{j,k} \quad \begin{matrix} 1 \le j \le N \\ 1 \le k \le M \end{matrix} \right\} \qquad (4.29)$$

Where, j indicates the column of pixels that can be varied from 1 to N, and k indicates the row of pixels varying from 1 to M.

The input video image contains the pixels. When the object detection is performed with the use of the pixels of the image, then the tracking accuracy gets reduced. The LBP feature (Dae-Hwan *et al.,* 2014) improves the tracking performance. The conventional LBP method utilizes the texture operator and thus describes each pixel of the small scale appearance of the input image using the thresholding technique (Soo-Chang *et al.,* 1999). The pixel of the image is formed against the value of the chosen centre pixel, and thus provides the binary value. Each neighbourhood value of the pixel can be formed by the equation 5.3. The binary value of 1 is provided for the condition when the value of the centre pixel is greater than the value of the centre pixel. The binary value is provided as 0 for the condition, when the neighbourhood value is lower than the centre pixel value.

$$S(x) = \left\{ \begin{matrix} 1 & if\ x > 0 \\ 0 & if\ x < 0 \end{matrix} \right\} \qquad (4.30)$$

equation 5.30 expresses the corresponding pixel value from the neighbourhood pixel. The corresponding pixel value is formed by the conversion of the binary values into decimal numbers.

$$v_i^L(j,k) = \sum_{a=0}^{7} S(i_a - i_c)2^a \qquad (4.31)$$

Where, i_c represents the centre pixel (j,k) represents the value of the eight surrounding pixels.

The LBP feature generates the new frame using equation 5.5.

$$V = \left\{ \begin{array}{ll} v_i^L(j,k) & 1 \le j \le N \\ & 1 \le k \le M \end{array} \right\} \qquad (4.32)$$

Finally, the two dimensional matrix is formed by combining both the original frame and the LBP generated frame as shown below:

$$f_{k,l} \in V_i, V_i^L \qquad (4.33)$$

$$F = \left\{ \begin{array}{ll} f_{k,l} & 1 \le K \le M \times N \\ & 1 \le l \le 2 \end{array} \right\} \qquad (4.34)$$

Where, V_i indicates the original frame,

V_i^L indicates the LBP generated frame and

$f_{k,l}$ indicates the combination of both the original frame and the generated LBP frame.

K-Means clustering

After obtaining the two dimensional matrix, the objects can be extracted using k-mean clustering, which considers the texture and other colour features to segment the objects. Then, group the data set with the aforementioned two dimensional matrix using the K means clustering algorithm. Based on the inherent distance from each other, the K-means clustering algorithm (Zhong *et al.* 2005) groups the input objects into multiple groups based on the features into K number of groups, where K is represented as any positive integer. The grouping done by the K-means clustering algorithm minimizes the sum of the squares of the distance between the object and the corresponding cluster centre. This is the type of partitioned clustering algorithm which determines all the clusters at once. The steps involved in the k means clustering algorithm are as follows:

i) Initialise the cluster centres in a two dimensional matrix based on the range of needed clusters C which can be represented as follows:

$$R = \begin{cases} R_{i,j} & 1 \leq i \leq C \\ & 1 \leq i \leq 2 \end{cases} \qquad (4.35)$$

ii) Randomly select the cluster centre $R_{i,j}$ in the two dimensional vector.

iii) The matching between every $f_{k,l}$ with R by the Euclidean distance, shows the similarity of the calculated two elements and influences the shape of the clusters.

$$ED(f_{k,l}, R) = \sqrt{\sum_{i=1}^{n} (f_{k,l} - R)^2} \qquad 0 \leq K \leq m \quad (4.36)$$

Where, $f_{k,l}$ is a two dimensional matrix with both the original frame and the LBP generated frame

iv) Find the minimum distance of $f_{k,l}$ with R, represented as $R_{new}(t)$

$$R_{new}(t) = \frac{1}{|C_1|} \sum_{f \in C_i} f_i^{k,l} \qquad \forall i \qquad (4.36)$$

v) Go to step (ii), until $R_{new}(t+1)$ is equal to $R_{new}(t)$

vi) Then, the number of objects selected based on the cluster value C, are shown as:

$R_1, R_2, ... R_C$

4.2.3.2 Visual Tracking Using the MOTSFK Second Derivative Model and Neighbourhood Search

Visual tracking (Heng and Ngan 1999) is a fundamental method for analysing video motion processing, data mining, visual surveillance, vision based control, and human computer interactions. The object appearance model of the visual tracking model is based on region colour histograms, kernel density estimates, Gaussian mixture model etc. However, multi-object visual tracking is still being researched because of a couple of factors, such as inter-acting each other posing inter-object occlusion, object confusion and low probability of detection. In order to overcome these aforementioned drawbacks in visual tracking, in this project the visual tracking algorithm based second derivative model and neighbourhood search, namely, SDVM (Second derivative visual model) MOTSFK is proposed.

Neighbourhood search algorithm

A Neighbourhood search algorithm (Kanungo and Qigong 2004) is one, in which the object can be tracked as follows: The fixed reference point in the template function is given in equation 5.37:

$$R_i^t \in f(x,y) \qquad (4.37)$$

Where, R_i^t indicates the reference point of the key frame. In the next process, each reference point from the new frame gets extracted based on the key frame reference point. Equation 5.12 shows the extracted reference point.

$$R_i^{t+1}(x,y) = f(r^k) \qquad (4.38)$$

where, R_i^{t+1} indicates the extracted reference point,

$f(r^k)$ indicates the function to generate a sub image.

The Euclidean distance between the key frame and the extracted frame provides the distance between the target objects in each frame and it is expressed by equation 5.13.

$$ED(R_i^t, R_i^{t+1}) = \sqrt{\sum_{i=1}^{n} \left(R_i^t - R_i^{t+1}(x,y)\right)^2} \qquad 0 \le K \le m \,(4.39)$$

The new location of the object in each frame is found by increasing or decreasing the reference point.

$$r_i^{t+1}(x,y) = \{x \pm i, y \pm j\} \quad (4.40)$$

Where, (i,j) indicates the increasing or decreasing reference point.

Now extract the new region from the input image and find the ED between the input image and the extracted region. The reference point with the minimum ED is taken as the final region of the detected image.

$$R_i^{t+1}(x,y) = \arg\left(\underset{K=1}{Min}\, P_K\right)(5.41)$$

MOTSFK matching formulae

The aforementioned neighbourhood search algorithm is unsuccessful to track the image if there is an interaction between multiple objects and frequent occlusions. In order to overcome these drawbacks, the second derivative model with the neighbourhood search algorithm can be utilized. The next spatial location of the object is found using second derivative-based visible pixels as follows:

$$P_K = \alpha \times ED\left(R_i^t, R_i^{t+1}(x,y)\right) + (1-\alpha) \times ED\left(\frac{\partial^2 R_i^t}{\partial x^2}, \frac{\partial^2 R_i^{t+1}(x,y)}{\partial x^2}\right)$$
$$+ (2-\alpha)ED\left(\frac{\partial^2 R_i^t}{\partial y^2}, \frac{\partial^2 R_i^{t+1}(x,y)}{\partial y^2}\right) \quad (4.42)$$

$$\frac{\partial^2 R_i^t}{\partial x^2} = \frac{\partial^2 f(x,y)}{\partial x^2} = W_{xx}(t)(4.43)$$

$$\frac{\partial^2 R_i^t}{\partial y^2} = \frac{\partial^2 f(x,y)}{\partial y^2} = W_{yy}(t)(4.44)$$

$$\frac{\partial^2 R_i^{t+1}(x,y)}{\partial x^2} = \frac{\partial^2 f(x,y)}{\partial x^2} = W_{xx}(t+1)(4.45)$$

$$\frac{\partial^2 R_i^{t+1}(x,y)}{\partial y^2} = \frac{\partial^2 f(x,y)}{\partial y^2} = W_{yy}(t+1)(4.46)$$

Where, R_i^t indicates the reference point in the key frame, $R_i^{t+1}(x,y)$ Indicates the reference point of the next frame,

$W_{xx}(t)$ and $W_{xx}(t+1)$ represent the second derivative of the key frame and the next frame based on x direction,

$W_{yy}(t)$ and $W_{yy}(t+1)$ represent the second derivative of the key frame and the next frame based on y direction.

ED represents the Euclidean distance between the key frame and the next frame of the reference point.

4.2.3.3 Spatial Tracking Using MOTSFK Tangent Weighted Spatial Tracking Model

The spatial tracking method (Kratz *et al.*, 2012), predicts the next spatial location of the object using the tangential weighted spatial model (TWSM). Here, the object is represented in the form of x, y coordinates in the key frame. According to the direction of the changing x, y coordinates, the corresponding object of the next frame can be predicted. Then, the location of the same object in the next frame is predicted according to the difference between the coordinates of the objects in the key frames with the next frame. By doing this tracking process using the spatial method, the object in the frame can be accurately tracked.

In this model, the centre point value of the objects is selected for further calculation. The selected centre point of the cluster can be represented as: $r_i^t .... r_C^t$ Where, r_i^t indicates the i-th object in the t-th frame. Using the aforementioned centre point of the object r_i^t, the spatial tracking method is performed using the following formula which is the EWMA model (Kanungo *et al.*, 2002).

$$tr_i^{t+1} = \alpha r_i^t + (1-\alpha) tr_i^t \quad (5.47)$$

Where, tr_i^{t+1} represents i-th object in the $t+1$-th frame. Then, the object r_i^t belongs to (i, j) Where, α represents a function of the exponential, in which each pixel intensity value of the output image is equal to the basis value raised to the value of the corresponding pixel values in the input image and represented as follows:

$$\alpha = \left(1 - \exp^{\Delta T / \tau}\right) \quad (4.48)$$

Where, τ represents the Time constant,

α represents the smoothing factor

ΔT represents the sampling time interval.

While doing this spatial tracking using the exponential function, several problems may occur during the tracking process. Therefore, to the best of our knowledge, in order to improve the tracking performance, this is the first ever research work where, the tangential weighted method is preferred instead of the exponential function.

Tangential weighed function

For improving the performance of spatial filtering, the tangential weighed function is preferred instead of the exponential function. The hyperbolic tangent is a function that can easily represent the transition from one state to another. The aforementioned tangential weighted function operates element wise on arrays, and has the advantage of non-linear and linear functions with high training speed. The prediction of the next location using the MOTSFK tangential function is shown as:

$$tr_i^{t+1} = \alpha r_i^t + (1 - \alpha) tr_i^t \quad (4.49)$$

$$\alpha = 2(1 - \tanh + C) \quad (4.50)$$

Where, r_i^{t+1} indicates the i-th object of the $t+1$-th frame

α indicates a smoothing factor,

tanh indicates the hyperbolic tangential function and

C indicates the weighted constant.

4.2.3.4 Integration of the Visual and Spatial Tracking Models

In this section, both the visual and spatial tracking models are integrated for better tracking performance. The results of the visibility model R_i^{t+1} and spatial tracking model tr_i^{t+1} are integrated to obtain the final tracking T_i^{t+1} results shown as:

$$T_i^{t+1} = \left[C\{R_i^{t+1}\} + \{tr_i^{t+1}\} \right] (4.51)$$

Where, R_i^{t+1} indicates a reference point selected in the visual representation,

$C\{R_i^{t+1}\}$ indicates a centre point of the object which has been selected from the reference area,

$\{tr_i^{t+1}\}$ indicates the centre point of the object in spatial tracking.

Fig 22 shows the algorithmic representation of multi-object tracking videos.

1	**Input:** Multi-object video, V
2	**Output:** Tracked object, T
3	**Parameters:** $R_i^{t+1} \rightarrow$ selected reference point in the visual representation
4	$tr_i^{t+1} \rightarrow$ centre point of the object in spatial tracking
5	$T_i^{t+1} \rightarrow$ Final tracking results
6	**Procedure**
7	**Begin**
8	Read the key frame from the video, V_i
9	Extract the LBP from the key frame
10	Generate two dimensional feature matrices $f_{k,l}$
11	Detect the objects from the video using the k-means clustering algorithm

12	For all the objects
13	Track the object using the formula of P_K in visual representation.
14	Find the cluster centre of the tracked object $C\{R_i^{t+1}\}$
15	Track the object using the spatial tracking method based on the formulae of tr_i^{t+1}
16	Integrate $T_i^{t+1} = [C\{R_i^{t+1}\} + \{tr_i^{t+1}\}]$
17	End for
18	End

Fig20: Algorithmic Description

4.2.3.5 Experimental Results

This section discusses the simulation results of the MOTSFK hybrid tracking model for tracking multiple target objects from the video frame. This section also compares the results of the MOTSFK hybrid tracking model with the existing second derivative visual method (SDVM), tangential weighed spatial model (TWSM) and the Gaussian mixture probability hypothesis density (GM-PHD) model. The performance metrics analyze the performance of the MOTSFK hybrid tracking model with the second derivative based visibility model and tangential weighted spatial tracking model.

4.2.3.5.1 Experimental Setup

The simulation of this research work is done using the Matlab 8.3 (R2014a) tool installed in the personal computer with the system configuration of 4 GB RAM Intel processor and a 64-bit operating system.

a) Dataset description

[1]. Videos with multiple objects are considered for the simulation of this research work. The UCSD datasets (http://www.svcl.ucsd.edu/projects/anomaly/dataset.htm.) has many kinds of videos with multiple objects. The MOTSFK model tracks the multiple objects from the video frame present in the UCSD dataset. The UCSD dataset contains videos of moving pedestrians or vehicles in a crowded area. The camera placed at the elevation records the movement of the various persons walking along the path. The path contains various objects, such as moving pedestrians, vehicles, bikers, carts etc. The density of the crowd varies over the time interval. The abnormalities in the video frame occur for these reasons :,

- The movement of objects in the video other than the pedestrians (pedestrians means people who are walking. No need to say 'walking pedestrians').
- The random movement of the pedestrians

The UCSD dataset contains two dataset video samples D1 and D2. The video dataset gets split based on the change in the background of the video. The samples in the subset contain an average of 200 video clip frames.

i) Data subset 1

The data subset 1 contains video samples of the group of pedestrians moving towards and away from the stationary camera. The video samples include the movement of the vehicles and bikes away and towards the camera. The data subset 1 contains a total of 34 training video samples and 36 testing video samples.

ii) Data subset 2

The data subset 2 contains video samples of the group of pedestrians moving parallel to the camera plane. The video samples include the movement of the vehicles and bikes parallel to the camera plane. The data subset 2 contains a total of 16 training video samples and 12 testing video samples.

The data subset from the UCSD dataset has varying videos for the simulation of the MOTSFK hybrid tracking model. The simulation of the research work makes use of four videos from the UCSD dataset. The first and second videos are taken from the data subset 1. The third and fourth videos are taken from the data subset 2. The videos taken have different crowd densities with multiple target objects. The UCSD dataset has video samples observed for a long interval of time with the ground truth information values. Since the UCSD dataset has the ground truth information, the performance of the MOTSFK hybrid tracking model can be analyzed through a comparison with the other existing models. The ground truth information is the actual measure of the target object movements in the video. The values obtained from the simulation of the MOTSFK hybrid model are compared with the ground truth information to measure the efficiency of the MOTSFK model. The performance of the hybrid model is compared with that of the existing models such as GM-PHD, TWSM, and the SDVM.

4.2.3.5.2 Evaluation Metrics for the Performance Analysis

The evaluation metrics measure the performance of the MOTSFK hybrid tracking model. This research employs three evaluation metrics for analyzing the hybrid tracking model. They are,

- Tracking Number
- Tracking Distance

- Multiple Object Tracking Precision (MOTP) (Ming *et al.* 2007).

a) Tracking Number

The target object to be tracked is present in many frames of the video. For the multiple-object tracking model, the target object presence in each frame varies with time. The video frames containing the target objects are given with the number in the increasing order. The tracking number defines the total number of frames in the video, which has the target object to be tracked. Equation 5.26 provides the expression for the tracking number.

Tracking Number = Number of frames in the video with the target object

5.26

b) Tracking Distance

The target objects present in the video vary their position and location in each frame. In The video, multiple target objects overlap each other during the tracking process. The position of the target objects needs to be found in each frame for better tracking. The tracking distance measures the distance of the pixel positions of the target object from the initial frame to the final frame. The tracking distance is the Euclidean distance between the frames containing the target object. The ground truth information contains the actual distance traveled by the target object from the initial frame to the final frame. The algorithm performance will be better if the tracking distance value is small. Equation 4.52 expresses the tracking distance for the video with n target objects.

$$Tracking\ Distance = \sqrt{\sum_{i=1}^{n} \left(R_i^{initial}(x,y) - R_i^{final}(x,y) \right)^2}$$

(4.52)

where, $R_i^{initial}$ indicates the initial frame with the target object

R_i^{final} indicates the final frame with the target object

(x, y) indicates the pixel position of the target object within the frame.

c) MOTP

The MOTP defines the measure of the tracking precision of the MOTSFK hybrid model. The ground truth information defines the position of the tracked target objects in each frame. The MOTP defines the error measure between the matched object-hypothesis pairs present in the frames of the video to the average number of matches made by the tracking model. The mathematical expression of the MOTP is given by equation 4.53. The MOTP value should be the maximum for better tracking.

$$MOTP = \frac{\sum_{i,t} D_t^i}{\sum_t m_t} \quad (4.53)$$

where, m_t represents the number of target object matches found at the time t, D_t^i represents the distance between the objects and target matching hypothesis in the video.

CHAPTER 5
RESULT

This section discusses the various experimental results of the MOTSFK hybrid tracking model for the four videos. The first video is taken from the data subset 1. The first video represents the movement of the pedestrians along with a bicycle. The first video has a crowded background, and hence, it allows the MOTSFK hybrid tracking model to track multiple target objects from the video frames. The first video gets split into 100 video clips of varying backgrounds. The input for the simulation process is chosen at random. 12 sample frames from video 1 are taken for the simulation. Video 1 shows the various pedestrians walking along the path (plate: 4).

Plate 4: Original Input Frames From Video 1

The frame samples from video 1 are subjected to k-means clustering for object detection in the video frames. The detected objects are surrounded by the yellow mark. K-means clustering performs the object detection by grouping the similar objects in the frame (plate:5).

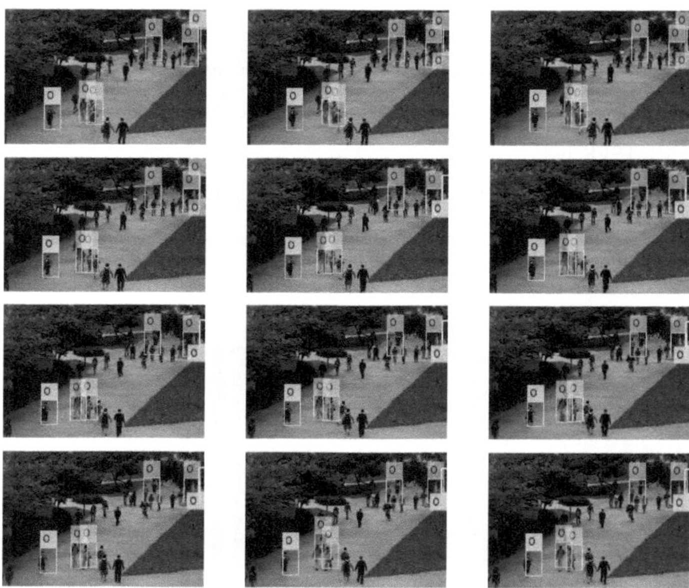

Plate 5: Detected Objects of The Frame From Video 1

The object tracking allows the user to predict the movements of the objects in every frame of the video. Object tracking is done after object detection in the frame. The Nearest neighbourhood algorithm performs the tracking of the detected objects in every frame of the video. The object tracking performance is done in various steps. If the detected objects overlap each other than the object tracking gets complicated. The tracked objects are marked with the blue mark (plate: 6).

Plate 6: Tracked Objects of The Frames From Video 1

The second video for the implementation of the MOTSFK model is taken from the data subset 1. The second video contains samples of the pedestrians in the path along with the moving car/van. The object occlusion is high for the second video. The movement of the people with the vehicle has high variation in the frame than the pedestrians. 12 sample frames from video 2 are taken for the simulation. Video 2 shows the various pedestrians walking on the path along with the movement of the bicycle and the van (plate: 7).

Plate 7: Original Input Frames From The Video 2

The frame samples from video 2 are subjected to the k-means clustering for object detection in the video frames. The detected objects are surrounded by the yellow mark. The clustering algorithm does not detect the van and the bicycle in the frame. Objects other than van/bicycle get detected by the algorithm (plate: 8).

Plate 8: Detected Objects of The Frame From Video 2

The resulted object tracked frame of video 2. The Nearest neighbourhood algorithm performs the tracking of the various objects in video 2. The tracked results from the figure show that the object positions in video 2 have small variations (plate: 9).

Plate 9: Tracked Objects of The Frame From Video 2

The third video for the implementation of the MOTSFK model is taken from the data subset 2. The third video contains samples of the pedestrians crossing the road with crowded density. The object occlusion is high for the third video; (plate: 10), shows 12 sample frames from video 3 taken for the simulation. These samples are continuous frames from video 3.

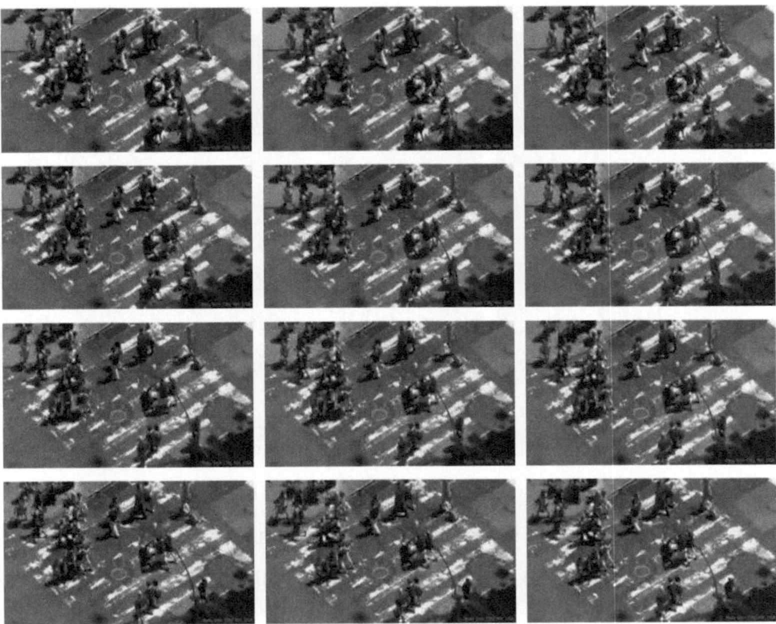

Plate 10: Original Input Frames From The Video 3

The frame samples from video 3 are subjected to the k-means clustering for object detection in the video frames. The detected objects in the frame of video 3. The detected objects are surrounded by the yellow mark. Since the crowd density is high in the third video, the target objects overlap each other. Thus, the objects detected in the third video frame sample are very less (plate: 11).

Plate 11: Detected Objects of The Frame From Video 3

The tracked object results of video 3 are shown. The tracking performed through the nearest neighbourhood algorithm has better results for this video, since the objects have less random movements during the road crossing. The object tracking in the video is directed more in the common direction (plate :11).

Plate 12 : Tracked Objects of the Frame from Video 3

The fourth video for the implementation of the MOTSFK model is taken from the data subset 2. The second video contains samples of the pedestrians in the halls/theatre. The object occlusion is very low for the fourth video. Video 4 contains fewer pedestrians than the other video samples (plate: 13).

Plate 13: Original Input Frames from Video 4

The frame samples from video 4 are subjected to the k-means clustering for object detection in the video frames. The detected objects are surrounded by the yellow mark. Video 4 has three target objects in each frame. The clustering algorithm detects every object in each frame (plate: 14).

Plate 14: Detected Objects of the Frame from Video 4

The object tracking results of video 4 are shown below. The three detected objects from the object detection are tracked at each frame of video 4. The aforementioned detected points from object detection allow the object tracking at each frame (plate: 15).

Plate 15: Tracked Objects of The Frame From Video 4

5.1 PERFORMANCE ANALYSIS OF THE HYBRID MODEL

The MOTSFK Hybrid tracking based on the second derivative based visibility model and tangential weighted spatial tracking model allows multi-object tracking in the videos. The performance of the MOTSFK model can be analysed by taking four videos from the UCSD dataset. The evaluation metrics, such as the tracking number, tracking distance, and the MOTP analyse the performance of the MOTSFK model. The performance of the MOTSFK model is compared with that of the existing algorithms, such as the Second Derivative Visual Method (SDVM), Tangential Weighed Spatial Model (TWSM), and the Gaussian Mixture Probability Hypothesis Density (GM-PHD) filter method.

5.1.1 Analysis Based on the Tracking Number

In this section, the performance of the models is compared based on the tracking number. The graph is drawn between the object number and the tracking number. Each detected object in the frame is given a number called as the object number. The performance of each model is compared with the ground truth information for each object number. The ground truth information provides the exact number of frames which has the target object. The model which has the tracking number closest to the ground truth information has the better result.

The performance of the MOTSFK hybrid model for the first video based on the tracking number is what?. The first video has a large number of objects, and hence, for simplification 8 objects are taken from each frame. The ground truth values for each of the object numbers of video 1 are 100, 100, 100, 100, 99, 95, 91, and 82 respectively. The SDVM model has the lowest performance based on the tracking number. For video 1, the existing SDVM tracks only 2 frames rather than the 100 frames for the object number 1. For the other object numbers, the SDVM shows the tracking number as 1. This indicates that the SDVM model has detected the objects in only one frame rather than 100 frames. The GM-PHD model shows the tracking numbers as 1, 4, 2, 2, 4, 6, 1,4 for object numbers from 1 to 8 respectively. This shows that the performance of the GM-PHD model is better than that of the SDVM model, but the variation between the tracking number of the GM-PHD and the ground truth information is large. The existing TWSM model has the tracking numbers of 98, 98, 98, 98, 97, 93, 89, 80 for the object numbers 1 to 8. The TWSM model attains a tracking number closer to the ground truth value. For video 1, the TWSM model detects 98 % of the target objects when compared to the other models. The MOTSFK hybrid model has the tracking numbers of 99, 99, 99, 99, 98, 94, 90, and 81 for the variation of the object number from 1 to 8. The value of the tracking number obtained by the MOTSFK hybrid model is most similar to the ground truth information rather than that of the other existing models. For video 1, the MOTSFK hybrid model detects 99 % of the target object frames when compared with the ground truth information.

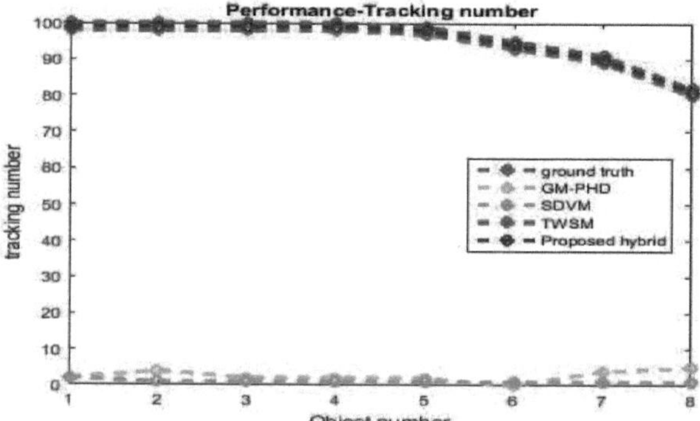

Fig 21: Performance Analysis For The MOTSFK Model With Video 1 Based on The Tracking Number

The performance of the MOTSFK hybrid model for the second video based on the tracking number. The second video has a large number of objects along with the movement of the van/bicycle, and hence for simplification 8 objects are taken from each frame. The ground truth values for each object numbers of video 2 are 71, 80, 81, 30, 30, 70, 77, and 71 respectively. The SDVM model has the lowest performance based on the tracking number. For video 2, the existing SDVM tracks only 2 frames rather than the 71 frames for the object number 1. For the other object numbers, the SDVM shows the tracking number as 1. This indicates that the SDVM model has detected the objects in only one frame rather than 80 frames. The GM-PHD model shows the tracking numbers as 1, 3, 1, 2, 1, 4, 1, and 2 for the object numbers from 1 to 8 respectively. The GM-PHD performance is better than that of the SDVM, but has a lower performance than the TWSM and the MOTSFK hybrid model. The existing TWSM model has the tracking numbers of 69, 78, 79, 28, 28, 68, 75, and 69 for the object numbers of 1 to 8. The TWSM model attains the tracking number closer to the ground truth value. For video 2, the TWSM model does not detect 2 frames with the target object when compared with the ground value. The MOTSFK hybrid model has the tracking numbers of 70, 79, 80, 29, 29, 69, 76, and 70 for the variation of the object numbers from 1 to 8. The value of the tracking number obtained by the MOTSFK hybrid model is most similar to the ground truth information rather than the other existing models. For video 2, the MOTSFK hybrid model detects 99% of the target object frames when compared with the ground truth information (Fig:21).

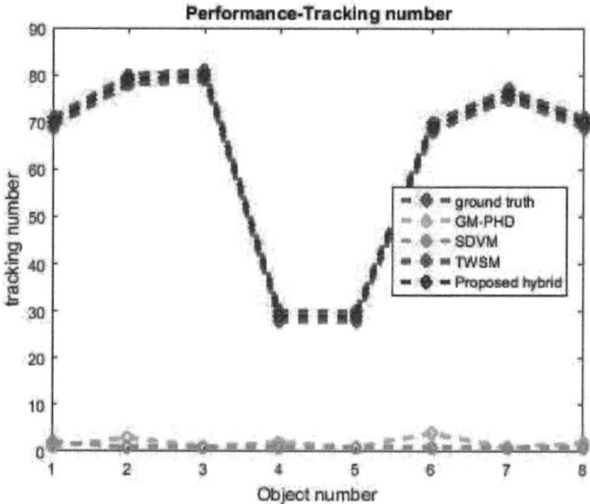

Fig 22: Performance Analysis For The MOTSFK Model With Video 2 Based On The Tracking Number

The performance of the MOTSFK hybrid model for the third video based on the tracking number. The third video has a large number of objects walking in the opposite directions, and hence, for simplification 8 objects are taken from each frame. The ground truth values for each object numbers of video 3 are 20, 72, 58, 44, 40, 83, 83, and 38 respectively. The SDVM model has the lowest performance based on the tracking number. For video 3, the existing SDVM has the tracking number as 2 for the object numbers 1 and 6. For the other object numbers, the SDVM shows the tracking number as 1. The SDVM has the maximum difference with the ground value compared to the other models. The GM-PHD model shows the tracking numbers as 2, 0, 2, 3, 1, 2, 4, and 2 for the object numbers from 1 to 8 respectively. This shows that the performance of the GM-PHD model is better than that of the SDVM model. For the object number 2, the GM-PHD model tracks any frame. The existing TWSM model has the tracking numbers of 18, 70, 56, 42, 38, 18, 70, and 56 for the object numbers = 1 to 8. The TWSM model attains the tracking number closer to the ground truth value for the object numbers 1 to 5. For other values of the object, it shows a major deviation in the track number. The MOTSFK hybrid model has the tracking numbers of 19, 71, 57, 43, 39, 19, 71, and 57 for the variation of the object numbers from 1 to 8. The value of the tracking number obtained by the MOTSFK hybrid model is most similar to the ground truth information rather than the other existing models. For video 3, the MOTSFK hybrid model detects 90 % of the target object frames higher than the SDVM, GM-PHD, and the TWSM models when compared with the ground truth information (Fig: 22).

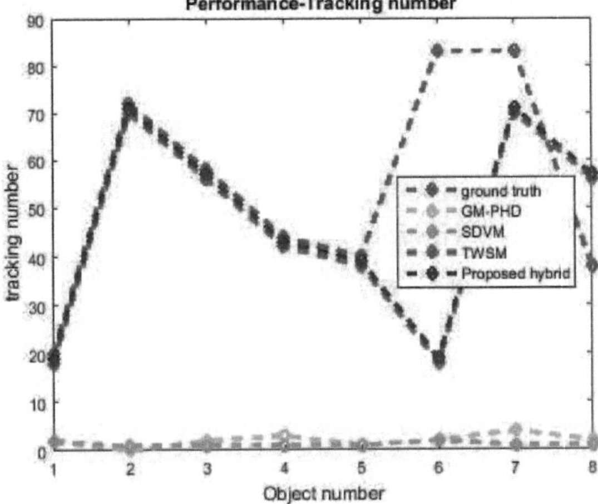

Fig 23: Performance Analysis For The MOTSFK Model With Video 3 Based on The Tracking Number

The performance of the MOTSFK hybrid model for the fourth video based on the tracking numbers is depicted in the graph. The fourth video has a small number of objects, and hence for simplification only 4 objects are taken from each frame rather than 8 objects. The ground truth values for each object numbers of video 4 are 100, 100, 100, and 100 respectively. The ground truth information indicates that each frame of the video contains every target object. The SDVM model has the lowest performance based on the tracking number. For video 4, the existing SDVM tracks only 1 frame rather than 100 frames for the object numbers of 1 to 4. This indicates that the SDVM model has detected the objects in only one frame rather than 100 frames. The GM-PHD model shows the tracking numbers as 2, 1, 3, and 5 for the object numbers from 1 to 4 respectively. This shows that the performance of the GM-PHD model is better than that of the SDVM model, but the variation between the tracking number of the GM-PHD and the ground truth information is large. The existing TWSM model has the tracking numbers of 98, 98, 98, and 98 for the object numbers of 1 to 4. The TWSM model attains the tracking number closer to the ground truth value. For video 4, the TWSM model detects 98 % of the target objects when compared to the other models. The MOTSFK hybrid model has the tracking numbers of 99, 99, 99, and 99 for the variation of the object numbers from 1 to 4. The value of the tracking number obtained by the MOTSFK hybrid model is most similar to the ground truth information rather than the other existing models. For video 4, the MOTSFK hybrid model detects 99 % of the target object frames when compared with the ground truth information (Fig: 23).

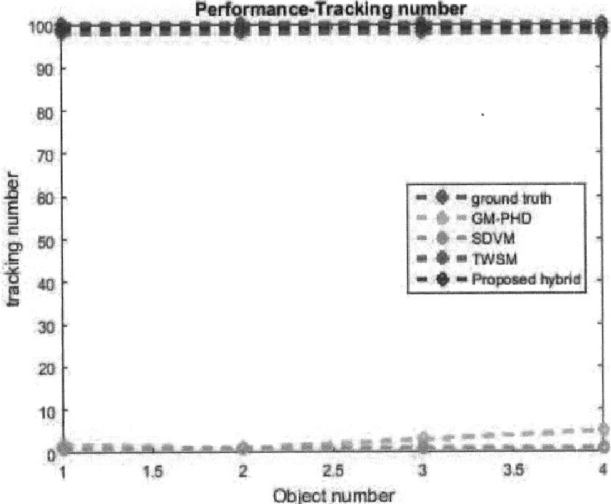

Fig 24: Performance Analysis For The MOTSFK Model With Video 4 Based On The Tracking Number

5.1.2 Analysis Based on the Tracking Distance

In this section, the performance of the models is compared based on the tracking distance. The graph is drawn between the object number and the tracking distance. The tracking distance measures the difference in the original path of the target object to the tracked path obtained by the algorithm. The distance moved by the target object in each frame is indicated by the tracking distance. The algorithm which provides the less difference value of the tracking distance has the better performance.

The performance of the MOTSFK hybrid model for the first video based on the tracking distance. The first video has a large number of objects, and hence, for simplification 8 objects are taken from each frame. The SDVM model has the lowest performance based on the tracking distance. For video 1, the existing SDVM has the tracking distance values of 1.3275, 1.4694, 1.4530, 1.6118, 1.9546, 2.0241, 1.9939, and 1.2964 for the object numbers of

1 to 8. When the object number is 7, the SDVM has the highest tracking distance measure. The difference in the tracking distance between the existing SDVM and the original frame is higher than that of the other models. The GM-PHD model shows the tracking distance as 1.343, 1.4625, 1.4557, 1.6076, 1.9301, 1.9793, 2.0023, and 1.2786 for the object numbers 1 to 8 respectively. This shows that the performance of the GM-PHD model is better than that of the SDVM model with lesser tracking distance variation. The existing TWSM model has the tracking distance of 0.0287, 0.0264, 0.0302, 0.0319, 0.0372, 0.0392, 0.0407, and 0.0308 for the object numbers of 1 to 8. The TWSM model has less variation in the tracking distance than the existing SDVM and the GM-PHD models. The MOTSFK hybrid model has the tracking distance of 0.0145, 0.0132, 0.0151, 0.0160, 0.0187, 0.0196, 0.0203, and 0.0155 for the variation of the object numbers from 1 to 8. The value of the tracking distance for the MOTSFK hybrid model is lesser than that of the existing models such as SDVM, GM-PHD, and the TWSM model (Fig: 24).

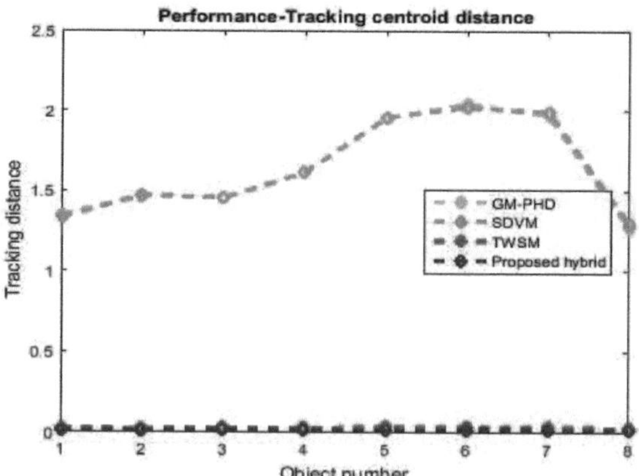

Fig 25: Performance Analysis For The MOTSFK Model With Video 1 Based On The Tracking Distance

The performance of the MOTSFK hybrid model for the second video based on the tracking distance. The SDVM model has the lowest performance based on the tracking distance. For video 2, the existing SDVM has the tracking distance values of 1.0152, 1.5945, 1.6942, 0.4671, 0.4926, 1.4927, 1.7139, and 1.1457 for the object numbers of 1 to 8. The difference in the tracking distance between the existing SDVM and the original frame is much higher than that of the other models. The GM-PHD model shows the tracking distance as 1.0367, 1.5864, 1.6960, 0.4708, 0.5013, 1.4732, 1.7171, and 1.1453 for the object numbers from 1 to 8 respectively. This shows that, the GM-PHD model has less tracking distance variation than the SDVM model. The existing TWSM model has the tracking distance of 0.0270, 0.0363, 0.0378, 0.0325, 0.0344, 0.0398, 0.0418, and 0.0355 for the object numbers of 1 to 8. The MOTSFK hybrid model has the tracking distance of 0.0135, 0.0181, 0.0189, 0.0162, 0.0173, 0.0199, 0.0209, and 0.0178 for the variation of the object numbers from 1 to 8. The value of the tracking distance for the MOTSFK hybrid model is lesser than that of the existing models such as SDVM, GM-PHD, and the TWSM model (Fig 25).

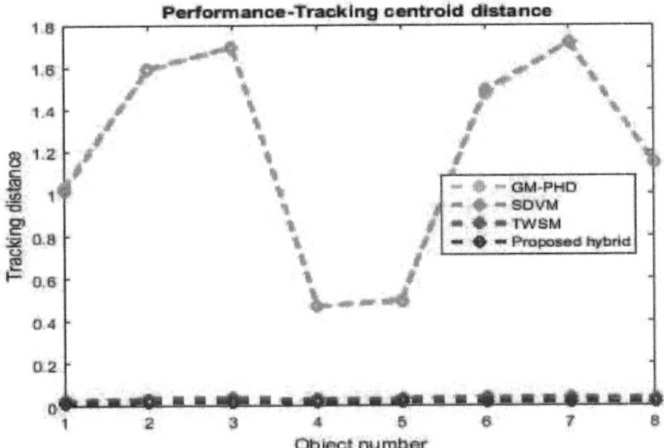

Fig 26: Performance Analysis For The MOTSFK Model With Video 2 Based On The Tracking Distance

The performance of the MOTSFK hybrid model for the third video based on the tracking distance is analysed. The third video has a large number of objects, and hence for simplification 8 objects are taken from each frame. The GM-PHD model has the lowest performance based on the tracking distance. For video 3, the existing GM-PHD has the tracking distance values of 0.0561, 0.9036, 0.8348, 0.8241, 0.4887, 1.6172, 1.6406, and 0.8631 for the object numbers of 1 to 8. The difference in the tracking distance between the existing SDVM and the original frame is higher than that of the other models. The SDVM model shows the tracking distance as 0.0493, 0.8961, 0.8376, 0.8192, 0.4777, 1.6220, 1.6731, and 0.8298 for the object numbers from 1 to 8 respectively. This shows that the performance of the SDVM model is better than that of the GM-PHD model with less tracking distance variation. The existing TWSM model has the tracking distance of 0.0079, 0.0150, 0.0205, 0.0299, 0.0314, 1.6529, 1.2944, and 1.6469 for the object numbers of 1 to 8. The TWSM model has less variation in the tracking distance for the object numbers 1 to 5 than the existing SDVM and the GM-PHD models. The MOTSFK hybrid model has the tracking distance of 0.0040, 0.0075, 0.0102, 0.0149, 0.0157, 1.6568, 1.3019, and 1.6571 for the variation of the object numbers from 1 to 8. The value of the tracking distance for the MOTSFK hybrid model is lesser than that of the existing models such as SDVM, GM-PHD, and the TWSM model for the lower object number in video 3(Fig: 26).

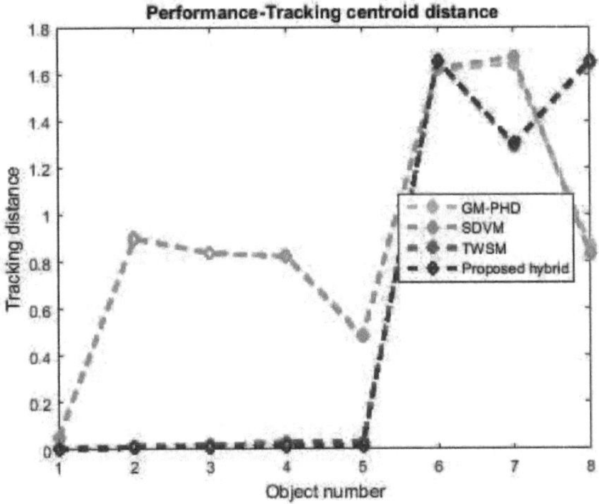

Fig 27: Performance Analysis For The MOTSFK Model With Video 3 Based
On The Tracking Distance

The performance of the MOTSFK hybrid model for the fourth video based
on the tracking distance is analysed. The fourth video has a small number of
objects, and hence, for simplification 4 objects are taken from each frame.
The SDVM model has the lowest performance based on the tracking distance.
For video 4, the existing SDVM has the tracking distance values of 5.1368,
8.4721, 8.3430, and 10.0813 for the object numbers of 1 to 4. When the object
number is 4, the SDVM has the highest tracking distance measure. The
difference in the tracking distance of the existing SDVM and the original
frame is higher than the other models'. The GM-PHD model shows the
tracking distance as 5.1183, 8.5148, 8.2985, and 9.9404 for the object numbers
from 1 to 4 respectively. This shows that the performance of the GM-PHD
model is better than that of the SDVM model with less tracking distance
variation. The existing TWSM model has the tracking distance 0.1027,
0.1782, 0.1667, and 0.1934 for the variation of the object numbers from 1 to 8.
The TWSM model has less variation in the tracking distance than the existing

SDVM and the GM-PHD models. The MOTSFK hybrid model has the tracking distance of 0.0513, 0.0885, 0.0837, and 0.0969 for the variation of the object numbers from 1 to 4. The value of the tracking distance for the MOTSFK hybrid model is lesser than that of the existing models such as the SDVM, GM-PHD, and the TWSM model (Fig: 27).

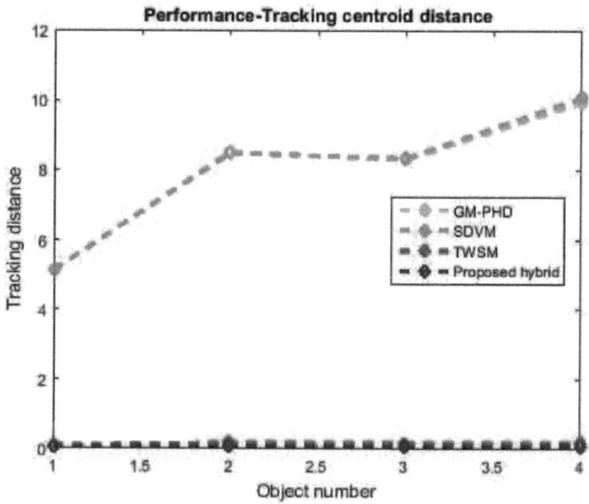

Fig28: Performance Analysis For The MOTSFK Model With Video 4 Based On The Tracking Distance

5.1.3 Analysis Based on MOTP

In this section, the performance of the models is compared based on MOTP. The graph is drawn between the object number and the variation in the MOTP metric. The algorithm with the highest MOTP value has the best performance.

The performance of the MOTSFK hybrid model for the first video based on the MOTP metric is analysed. The first video has large number of objects, and hence for simplification 8 objects are taken from each frame. For video 1, the existing GM-PHD has MOTP values 0, 0, 0, 0, 0.0100, 0.0500, 0.0900, and 0.1800 for the object numbers of 1 to 8. The GM-PHD model has a low value of MOTP for the object numbers 1 to 4. The SDVM model shows the MOTP as 0.0100, 0.0100, 0.0100, 0.0100, 0.0200, 0.0600, 0.1000, and 0.1900 for the object numbers from 1 to 8 respectively. The SDVM model has the better MOTP of 6% at the object number 6. The existing TWSM model has the MOTP of 0.9800, 0.9800, 0.9800, 0.9800, 0.9800, 0.9800, 0.9800, and 0.9800 for the object numbers of 1 to 8. The MOTP of the TWSM model has a better value of 98 % for all the object numbers for the first video. The MOTSFK hybrid model has the MOTP of 0.9900, 0.9900, 0.9900, 0.9900, 0.9900, 0.9900, 0.9900, and 0.9900 for the variation of the object numbers from 1 to 8. The MOTSFK hybrid model has the best MOTP value of 99 % for each object number and thus overcomes the existing models in the MOTP performance(Fig:28).

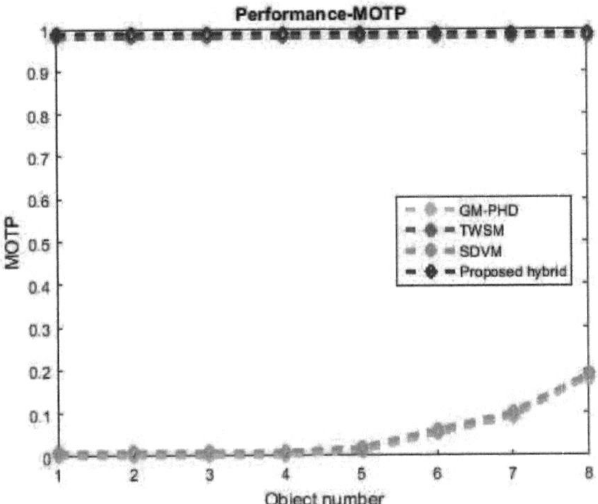

Fig 29: Performance Analysis For The MOTSFK Model With Video 1 Based On The MOTP Parameter

The performance of the MOTSFK hybrid model for the second video based on the MOTP metric. The second video has a large number of objects, and hence, for simplification 8 objects are taken from each frame. For video 2, the existing GM-PHD has the MOTP values 0.2900, 0.2000, 0.1950, 0.6950, 0.7000, 0.3000, 0.2300, and 0.2900 for the object numbers of 1 to 8. The SDVM model shows the MOTP as 0.3050, 0.2100, 0.2050, 0.7100, 0.7100, 0.3100, 0.2400, and 0.3000 for the object numbers from 1 to 8 respectively. The existing GM-PHD and the SDVM models have better MOTP for the object numbers of 4 and 5. But, for other values the MOTP value is lower. The existing TWSM model has the MOTP of 0.9800, 0.9800, 0.9800, 0.9800, 0.9800, 0.9800, 0.9800, and 0.9800 for the object numbers of 1 to 8. The existing TWSM model has the improved MOTP of 98 % and thus is better than SDVM and the GM-PHD model. The MOTSFK hybrid model has the MOTP of 0.9900, 0.9900, 0.9900, 0.9900, 0.9900, 0.9900, 0.9900, and 0.9900 for the variation of the object numbers from 1 to 8. The MOTSFK hybrid model has the best MOTP value of 99 % for all object numbers (Fig: 29).

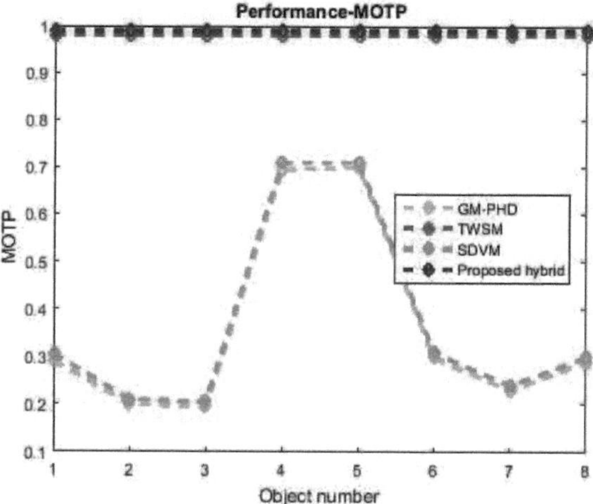

Fig 30: Performance Analysis For The MOTSFK Model With Video 2 Based On The MOTP Parameter

The performance of the MOTSFK hybrid model for the third video based on the MOTP metric. The third video has a large number of objects, and hence for simplification 8 objects are taken from each frame. For video 3, the existing GM-PHD has the lowest MOTP values of 0.7900, 0.2800, 0.4200, 0.5500, 0.5900, 0.1550, 0.1700, and 0.5850 for the object numbers of 1 to 8. The GM-PHD has low MOTP performance compared to the other models. The SDVM model shows the MOTP as 0.8100, 0.2900, 0.4300, 0.5700, 0.6100, 0.1500, 0.1600, and 0.6100 for the object numbers from 1 to 8 respectively. The SDVM model has the MOTP of 81% when the object number is 1. The existing TWSM model has the MOTP of 0.9800, 0.9800, 0.9800, 0.9800, 0.9800, 0.0200, 0.0250, and 0.0600 for the object numbers of 1 to 8. The TWSM model has higher MOTP for object numbers 1 to 5, but for higher object number, the TWSM model is not suitable. The MOTSFK hybrid model has the MOTP of 0.9900, 0.9900, 0.9900, 0.9900, 0.9900, 0.0100, 0.0150, and 0.0500 for the variation of the object numbers from 1 to 8. The MOTSFK hybrid model outperforms other existing models with the MOTP of 99 % for the lower object numbers (Fig: 30).

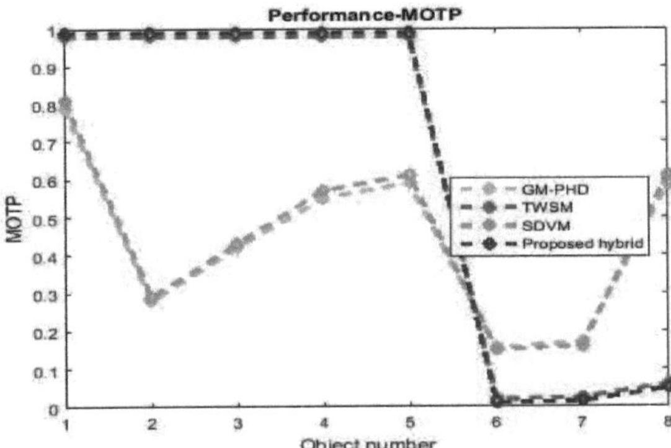

Fig 31: Performance Analysis For The MOTSFK Model With Video 3 Based On The MOTP Parameter

The performance of the MOTSFK hybrid model for the fourth video based on the MOTP metric is analysed. The fourth video has a large number of objects, and hence for simplification 4 objects are taken from each frame. For video 4, the existing GM-PHD has the MOTP value 0, 0, 0, and 0 for the object numbers of 1 to 4. The SDVM model shows the MOTP as 0.0100, 0.0100, 0.0100, and 0.0100 for the object numbers from 1 to 4 respectively. The existing TWSM model has the MOTP of 0.9800, 0.9800, 0.9800, and 0.9800 for the object numbers of 1 to 4. The MOTSFK hybrid model has the MOTP of 0.9900, 0.9900, 0.9900, and 0.9900 for the variation of the object numbers from 1 to 4. The results show that the MOTSFK hybrid model has the overall better performance than the other existing techniques (Fig: 31).

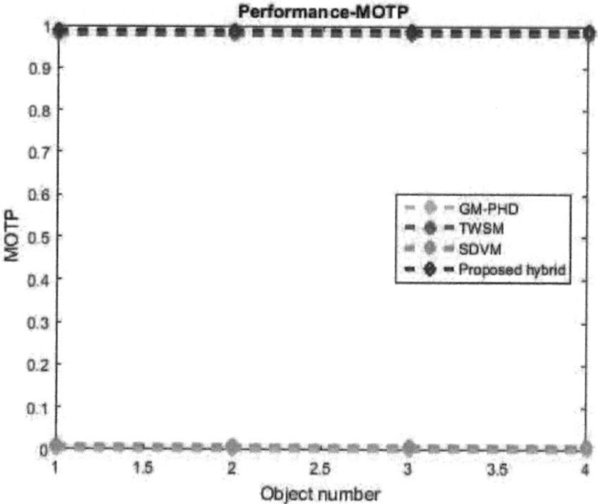

Fig 32: Performance Analysis For The MOTSFK Model With Video 4 Based On The MOTP Parameter

This section explains the results of the MOTSFK model with its comparison with the existing models such as GM-PHD, SDVM, and TWSM. The comparison is done based on the performance metrics, such as the tracking number, tracking distance, and the MOTP. The comparison is done for the four videos from the database.

Table 5.4 shows the comparison results of the MOTSFK hybrid tracking model with the existing models based on the tracking number. The performance of each tracking model based on the tracking number can be measured by comparing the tracking number values of the models with the ground value information. The ground value information provides the exact number of the target objects in the frame of the video. When the ground value of video 1 is 100, then the existing GM-PHD model has the lowest tracking number of 2. The existing SDVM has the better performance than the GM-PHD with the tracking number value of 6. But, the values of the tracking number of the existing SDVM and the GM-PHD show that these models have very low performance. The existing TWSM model tracks 98% of the frames with the target object for video 1 which is significantly better than the existing SDVM and the GM-PHD models. The MOTSFK hybrid tracking model tracks 99 frames out of 100 frames with the target object for video 1. For video 2, when the ground value is 81 the existing GM-PHD model and the SDVM model have the lowest tracking number of 1. The existing TWSM model tracks 98% of the frames with the target object for video 2 which is significantly better than the existing SDVM and the GM-PHD models. The TWSM model has the tracking number of 79 out of 81 as the ground value. The MOTSFK hybrid tracking model tracks 80 frames out of 81 frames with the target object for video 2.

When the ground value of video 3 is 58, then the existing GM-PHD model has the lowest tracking number of 1. The existing SDVM has the better performance than the GM-PHD with the tracking number value of 2. The existing TWSM model tracks 98% of the frames with the target object for video 3 which is significantly better than the existing SDVM and the GM-PHD models. The MOTSFK hybrid tracking model tracks 57 frames out of 58 frames with the target object for video 3. For video 4, when the ground value is 100 the existing GM-PHD model and the SDVM model have the lowest tracking numbers of 1 and 2 respectively. The existing TWSM model tracks 98% of the frames with the target object for video 4, which is significantly better than the existing SDVM and the GM-PHD models. The MOTSFK hybrid tracking model tracks 99 frames out of 100 frames with the target object for video 4.

Table 3 : Comparison of The MOTSFK Model With The Existing Models Based on The Tracking Number

Videos	Ground value	Tracking Number			
		GM-PHD	SDVM	TWSM	MOTSFK hybrid tracking model
Video 1	100	2	6	98	99
Video 2	81	1	1	79	80
Video 3	58	1	2	56	57
Video 4	100	1	2	98	99

Table 3 shows the comparison results of the MOTSFK hybrid tracking model with the existing models based on the tracking distance. The tracking distance measures the movement of the target object within the frame. The value of the tracking distance should be low to get a more efficient model. The existing SDVM model has the highest tracking distance of 1.4694. The existing GM-PHD model has the better performance than the SDVM with the tracking distance value of 1.4625. The existing TWSM model has better tracking distance of 0.0264 than the existing SDVM and the GM-PHD models. The MOTSFK hybrid tracking model has the lowest tracking distance value of 0.013 for video 1 outperforming the existing models. For video 2, the existing GM-PHD model and the SDVM model have the lowest tracking distance of 1.0367 and 1.0152 respectively. The existing TWSM model has the tracking distance value as 0.0270 which is significantly lower than the GM-PHD and the SDVM models. The MOTSFK hybrid tracking model outperforms the existing models such as SDVM, GM-PHD, and TWSM with the lowest tracking distance value of 0.0135 for video 2.

For video 3, the existing GM-PHD model has the highest tracking distance of 0.0561. The existing SDVM has the better tracking distance than the GM-PHD with the value of 0.0493. The existing TWSM model has the tracking distance value of 0.0079. The MOTSFK hybrid tracking model has the overall best performance with the lowest tracking distance value of 0.0040 for video 3. For video 4, the existing SDVM model has the highest tracking distance of 5.1368. The existing GM-PHD has a better tracking distance than the SDVM with the value of 5.1183. The existing TWSM model has the tracking distance value of 0.1027 which is lower than the GM-PHD and the SDVM. The MOTSFK hybrid tracking model outperforms the existing models such as SDVM, GM-PHD, and TWSM with the lowest tracking distance value of 0.0513 for video 4.

Table 4: Comparison of The MOTSFK Model With The Existing Models Based on The Tracking Distance

Videos	Tracking Distance			
	GM-PHD	SDVM	TWSM	MOTSFK hybrid tracking model
Video 1	1.4625	1.4694	0.0264	0.0132
Video 2	1.0367	1.0152	0.0270	0.0135
Video 3	0.0561	0.0493	0.0079	0.0040
Video 4	5.1183	5.1368	0.1027	0.0513

Table 4 shows the comparison results of MOTSFK hybrid tracking model with the existing models based on the MOTP values. The MOTP defines the measure of the tracking precision of each model. The MOTP defines the error measure between the matched object-hypothesis pairs present in the frames of the video to the average number of matches made by the tracking model. The model with the highest MOTP value has the overall best performance.

For video 1, the existing GM-PHD model has the lowest MOTP value of 0.1800. The existing SDVM model has a better performance than the GM-PHD with the MOTP value of 0.1900. The existing TWSM model has a better MOTP value of 0.9800 than the existing SDVM and the GM-PHD models. The MOTSFK hybrid tracking model has the MOTP value of 0.9900 for video 1 outperforming the existing models. For video 2, the existing GM-PHD model and the SDVM model have the lowest MOTP values of 0.7000 and 0.7100 respectively. The existing TWSM model has the MOTP value as 0.9800 which is significantly higher than the GM-PHD and the SDVM models. The MOTSFK hybrid tracking model outperforms the existing models such as SDVM, GM-PHD, and TWSM with the MOTP value of 0.9900 for video 2.

For video 3, the existing GM-PHD model has the MOTP value of 0.7900. The existing SDVM has the better MOTP value than the GM-PHD with the value of 0.8100. The existing TWSM model has the MOTP value of 0.9800. The MOTSFK hybrid tracking model has the overall best performance with the MOTP value of 0.9900 for video 3. For video 4, the existing SDVM model has the MOTP value of 0.100. The existing GM-PHD has the MOTP value of 0. The existing TWSM model has the MOTP value of 0.9800 which is higher than the GM-PHD and the SDVM. The MOTSFK hybrid tracking model outperforms the existing models such as SDVM, GM-PHD, and TWSM with the MOTP value of 0.9900 for video 4.

Table 5: Comparison of The MOTSFK Model With The Existing Models Based on MOTP

Videos	MOTP			
	GM-PHD	SDVM	TWSM	MOTSFK hybrid tracking model
Video 1	0.1800	0.1900	0.9800	0.9900
Video 2	0.7000	0.7100	0.9800	0.9900
Video 3	0.7900	0.8100	0.9800	0.9900
Video 4	0	0.0100	0.9800	0.9900

CHAPTER 6
DISCUSSION

From the above results, it is clear that the MOTSFK hybrid tracking model has a better performance than the existing models, such as Gaussian mixture probability hypothesis density (GM-PHD) filter (Xiaolong *et al.*, 2014), SDVM, and TWSM. Video 1 contains more number of target objects. The tracking distance of the MOTSFK model is the minimum for video 1 than the other videos. Also, the MOTSFK model tracks 99 % of the target objects in each video, which is clearly better than that of the existing models. The MOTP measure of the MOTSFK model is 99 % for the four videos. This proves that the MOTSFK hybrid tracking model outperforms the existing models for the metrics of tracking distance, tracking number, and MOTP.

In the method for object detection and tracking of the objects in the video using the two techniques, namely, object segmentation and feature extraction process. The visual tracking method of object detection and tracking comprises of two modules, namely, the object detection module and the object tracking module. In the object detection module, the location of the object is in the video. The Shot segmentation process is performed in this detection module, and initially, in this section, the videos are converted into frames. The location of the object in each frame is found and the position is determined. The cropped image from the first frame is placed as the reference object and this reference object is used to detect the object in the other frames for enabling perfect object tracking. Experimentation is performed with different videos like the car1 video, car2 video, face video, and the akiyo video. The performance analysis is carried out in terms of two parameters like the error value and the score value. The error value should be a low one and the score value should be the maximum for an effective method. The comparison of the performance parameters of visual tracking in terms of the existing methods proves that the visual tracking method attained a minimum error value and the highest score value of 0.93 when compared with the existing methods.

This chapter also presents the multiple object detection and tracking method using the MOTSFK shape-based features and the Kalman filter. Here, there are two phases, namely, the object detection phase and the object tracking phase. In the object detection phase, there are two processes, namely, shot segmentation and manual segmentation. Initially, the video is converted into frames and the corresponding frame is selected using the parameter. The parameter used for selecting the frame is the distance metric. The frame with the minimum distance measure is selected and then the object is located in the frame. This object is located such that it satisfies the minimal area. Once the object is located, determine the centre point of the object and localize the object in all the frames. The above-mentioned procedure is repeated for perfect object tracking and then the particular object is tracked in the video. For object tracking, the centroid of the object is calculated using the Kalman filter. If the centroid calculated using the Kalman filter is the same as the centroid of the object, then locate the position of the object to be the same;, otherwise update the position of the object. The visual tracking method is capable of tracking multiple objects from the video. The effectiveness of the visual tracking method is proved using the two metrics, namely, the error and the score values. The visual tracking method proved to have the minimum error value and the maximum score value of 1.8 and 0.95 respectively, which proves the effectiveness of the visual tracking method.

This research work also proposes a hybrid tracking model for tracking multiple targets from the video. The MOTSFK hybrid tracking model uses the visual tracking and the spatial tracking techniques to track the target objects. The MOTSFK hybrid tracking model has a twofold architecture. The visual tracking model used in this work is a second derivative model. The spatial tracking model is a tangential weighted model. The MOTSFK model tracks the multiple objects from the video frame present in the UCSD dataset. The simulation results show that the MOTSFK hybrid tracking model has the improved performance than the existing models such as GM-PHD, SDVM, and TWSM. The tracking distance of the MOTSFK model is 0.0132 which is the minimum among the existing models. Also, the MOTSFK model tracks 99 % of the target objects in each video, which is clearly better than that of the existing models. The MOTP measure of the MOTSFK model is 99 % for the four videos. This proves that the MOTSFK hybrid tracking model outperforms the existing models for the metrics of tracking distance, tracking number, and the MOTP.

CHAPTER 7

SUMMARY

The method for object detection and tracking of the objects in the video using the two techniques, namely the object segmentation and the feature extraction process. The visual tracking method of object detection and tracking comprises of two modules, namely, the object detection module and the object tracking module. In the object detection module, the location of the object is found in the video. Shot segmentation is performed in this detection module, and initially, in this section, the videos are converted into frames. The location of the object in each frame is located and the position is determined. The cropped image from the first frame is placed as the reference object and this reference object is used to detect the object in the other frames for enabling perfect object tracking. Experimentation is performed with different videos like the car1 video, car2 video, face video, and the akiyo video. The performance analysis is carried out in terms of two parameters like the error value and the score value. The error value should be a low one and the score value should be the maximum value for an effective method. The comparison of the performance parameters of visual tracking in terms of the existing methods proves that the visual tracking method attained a minimum error value and the highest score value of 0.93 when compared with the existing methods.

This chapter also presents the multiple object detection and tracking method using the MOTSFK shape-based features and the Kalman filter. Here, there are two phases, namely, the object detection phase and the object tracking phase. In the object detection phase, there are two processes, namely, shot segmentation and manual segmentation. Initially, the video is converted into frames and the corresponding frame is selected using the parameter. The parameter used for selecting the frame is the distance metric. The frame with the minimum distance measure is selected and then the object is located in the frame. This object is located such that it satisfies the minimal area. Once the object is located, determine the centre point of the object and localize the object in all the frames. The above-mentioned procedure is repeated for perfect object tracking and then the particular object is tracked in the video. For object tracking, the centroid of the object is calculated using the Kalman filter. If the centroid calculated using the Kalman filter is the same as the centroid of the object, and then locate the position of the object to be the same; otherwise update the position of the object. The visual tracking method is capable of tracking multiple objects from the video. The effectiveness of the visual tracking method is proved using the two metrics, namely, the error and the score values. The visual tracking method proved to have the minimum error value and the maximum score value of 1.8 and 0.95 respectively, which proves the effectiveness of the visual tracking method.

This research work also proposes a hybrid tracking model for tracking multiple targets from the video. The MOTSFK hybrid tracking model uses the visual tracking and the spatial tracking techniques to track the target objects. The MOTSFK hybrid tracking model has a twofold architecture. The visual tracking model used in this work is a second derivative model. The spatial tracking model is a tangential weighted model. The MOTSFK model tracks multiple objects from the video frame present in the UCSD dataset. The simulation results show that the MOTSFK hybrid tracking model has a better performance than the existing models such as GM-PHD, SDVM, and TWSM. The tracking distance of the MOTSFK model is 0.0132 which is the minimum of the existing models. Also, the MOTSFK model tracks 99 % of the target objects in each video which is clearly better than that of the existing models. The MOTP measure of the MOTSFK model is 99 % for the four videos. This proves that the MOTSFK hybrid tracking model outperforms the existing models for the metrics of tracking distance, tracking number, and the MOTP.

CHAPTER 8
FUTURE DIRECTION

The future direction of the multi-object tracking model islisted as follows:,

> ➢ In real-time surveillance, if more than one camera is present then the target objects are tracked in any one of the cameras. This increases the tracking complexity.

> ➢ If the camera is mobile, then more features need to be included in the multi-object tracking process.

> ➢ Due to the improvement in computational power, computers are needed to provide visual capacities. Now –a- days, there is a rapid development in the field of vision-based interfaces and video surveillance applications where visual tracking is significant.. The objective of visual tracking is to generate the non-invasive and natural human computer interfaces and visually identify the actions of human beings. Many research works have been devoted to video analysis and the visual tracking problem. The major objective of visual tracking is finding the target state from a given video input. Visual tracking techniques are also used to recover states such as 2D positions, articulations, and poses. The target states are concealed in an image sequence and gathered by the visual features.

> ➢ The following work can be used in high security areas where continuous detection and tracking is used.

CHAPTER 9

CONCLUSION

This chapter concludes this research work with a summary, major findings, and future work. This research is aimed at revealing the importance and challenges of multi-object tracking in a video. This research has effectively analysed the difficulties in multi- object tracking in videos and has developed various models to track multiple target objects from videos. The primary contribution of this research work is multi-object tracking with the shot segmentation model with the extracted features, such as the wavelet moment, histogram, shape, and cologram. In the second part, multiple object tracking was performed with the shape-based features such as the centroid and tilt and tracking through the Kalman filter. In the third work, a hybrid tracking model based on the visual and spatial tracking MOTSFK was proposed. The performance metrics, are error value, score value, tracking number, tracking distance, and the MOTP. The simulation of the MOTSFK work is done by taking various video samples from the UCSD dataset. The performance of the research work is analyzed by comparing it with the existing models such as GM-PHD, SDVM, and TWSM. The simulation results show that the MOTSFK wavelet based object tracking method achieved a minimum error performance and the score value of 0.93. The MOTSFK Kalman filter multi-object tracking method has the maximum score values of 0.95 and minimum error value of 1.8. For the MOTSFK hybrid tracking model based on spatial and visual tracking, the simulation results show that the MOTSFK model has achieved a minimum tracking distance of 0.0132 and the MOTSFK model tracks 99 % of the target objects in each video. The MOTP measure of the MOTSFK hybrid tracking model is 99 % for each video. From the simulation results, it is evident that the MOTSFK work has outperformed the existing models in each performance metric. Since, the videos used have different crowd densities, the MOTSFK model from this research work can be utilized for video surveillance in all environments.

BIBLOGRAPHY

1. Abhineet, S. (2016). Robust Tracking for Multiple Objects in a video.*in proceedings of3rd International Conference on Computing for Sustainable Global Development (INDIACom).*

2. Ali, A. and Aggarwal, J. (2001). Segmentation and recognition of continuous human activity.*IEEE Workshop on Detection and Recognition of Events in Video.,28–35.*

3. Alper, Y. Omar, J. and Mubarak, S. (2006). Object Tracking A Survey.*ACM Computing Surveys.,38(4):303-312.*

4. Amit, Kumar, K,C. Laurent, J. and Christophe, De, Vleeschouwer. (2017) Discriminative and Efficient Label Propagation on Complementary Graphs for Multi-Object Tracking.*IEEE Transactions on Pattern Analysis and Machine Intelligence.,39(1) : 61 - 74.*

5. Badri N. S., Pradipta K. N., and Ashish G. (2011).A Change Information Based Fast Algorithm for Video Object Detection and Tracking.*IEEE Transactions on Circuits and Systems for Video Technology.,21(7) : 993-1004.*

6. Ballard, D. and Brown, C. Computer Vision, Prentice-Hall, 1982.

7. Beyan, C. and Temizel, A. (2012). Adaptive mean-shift for automated multi object tracking. *IET Computer Vision., 6(1) : 1 – 12.*

8. Bhanu, B. and Burger, W. (1990) .Qualitative understanding of scene dynamics for moving robots. *International Journal of Robotics Research.,9(6) : 74-90.*

9. Black, J. Ellis, T. and Rosin, P. (2002).Multi-View Image Surveillance and Tracking. *IEEE Workshop on Motion and Video Computing.*

10. Black, M. and Jepson, A. (1998). Eigen tracking: Robust matching and tracking of articulated objects using a view-based representation, *International Journal of Computer Vision., 26(1) : 63–84.*

11. Breitenstein, M,D. Reichlin, F. Leibe, B. Meier, E, K. and Gool, L,V. (2009).Robust tracking-by-detection using a detector confidence particle filter., *Computer Vision IEEE 12th International Conference.*

12. Chenglizhao, C. Shuai, L. Hong, Q. and Aimin, H. (2015). Real-time and Robust Object Tracking in Video via Low-Rank Coherency Analysis in Feature Space. *Pattern Recognition.,* 48(9) : 2885–2905.

13. Comaniciu, D.Ramesh,V. and Meer, P. (2003). Kernel-based object tracking,*IEEE Transaction on Pattern Analysis and Machine Intelligence.,* 25(5) : 564–577.

14. Comport, A. Marchand, E. Pressigout, M. and Chaumette, F. (2006). Real-time markerless tracking for augmented reality: the virtual visual serving frame work. *IEEE Trans. Vis. Comput. Graph.,*12(4) : 615–628.

15. Cremers, D. (2006).Dynamical statistical shape priors for level set based tracking. *IEEE Transaction on Patten Analysis and Machine Intelligence.,* 28(8) : 1262–1273.

16. Cremers, D. Rousson, M. and Deriche, R. (2007). A review of statistical approaches to level set segmentation: Integrating color, texture, motion and shape. *International Journal of Computer Vision.,* 72(2) : 195–215.

17. Dae-Hwan, K. Hyo-Kak, K. Seung-Jun, L. Won-Jae, P. and Sung-Jea K(2014).Kernel-Based Structural Binary Pattern Tracking. *IEEE Transactions on Circuits and Systems for Video Technology.,* 24(8) : 1288-1300.

18. Debi, P,D. Arun, K, M. Shamik, S. Jayanta, M. Suchandra, M. and Arun, S. (2012). Toward Automating Hammersmith Pulled-To-Sit Examination of Infants Using Feature Point Based Video Object Tracking, *IEEE Transactions on Neural Systems and Rehabilitation Engineering.,* 20(1) : 38-47.

19. Denis, K. Garik, M. and Dmitry, K. (2015).Data Fusion for Unsupervised Video Object Detection, Tracking and Geo-Positioning. *proceedings of 18th International Conference on Information Fusion (Fusion).*

20. Dorra, R. and Guillaume-Alexandre, B. (2016). Online multi-object tracking by detection based on generative appearance models. *Computer Vision and Image Understanding.,* 152(1) : 88–102.

21. Edwards, G. Taylor, C. and Cootes, T. (1998). Interpreting face images using active appearance models. *Proceedings of International Conference on Face and Gesture Recognition.,* 300–305.

22. Eiselein, V. Arp, D. Pätzold, M. Sikora, T. (2012).Real-time multi-human tracking using a probability hypothesis density filter and multiple detectors.*in: AVSS.*

23. Faegheh, S. and Mohsen, E, M. (2016). An object tracking method using modified galaxy-based search algorithm. *Swarm and Evolutionary Computation.,* 30(1) : 27–38.

24. Feng, X. (2016). Research on PSOGA Particle Filter Video Object Tracking Algorithm Based on Local Multi-zone, *Control and Decision Conference (CCDC).*

25. Fieguth, P. and Terzopoulos, D. (1997). Color-based tracking of heads and other mobile objects at video frame rates. *In proceedings of IEEE Conference on Computer Vision and Pattern Recognition (CVPR).* 21–27.

26. Frey, B, J. Jojic, N. and Kannan, A. (2003). Learning appearance and transparency manifolds of occluded objects in layers. *In Proceedings of IEEE Conference on Computer Vision and Pattern Recognition.,* 1(1) : 45–52.

27. Fussenegger, M. Roth, P, M. Bischof, H. and Pinz, A. (2006). Online, incremental learning of a robust active shape model.*in Proceedings of DAGM-Symposium on Pattern Recognition.,*122–131.

28. Guang, H. Xingyue, W. Jixin L, Ning, S. and Cailing, W. (2016). Robust object tracking based on local region sparse appearance model.*Neurocomputing,* 184(1) : 145–167.

29. Guillaume, C. Amaury, D. and Eric, M. (2014).Direct model based visual tracking and pose estimation using mutual information. *Image and Vision Computing, Image and Vision Computing.,*32(1) : 54–63.

30. Guoheng, H. Chi-Man, P. and Cong, L. (2015). Video Object Tracking Using Interactive Segmentation and Super pixel Based Gaussian Kernel. In *proceedings of 19th International Conference on Information Visualisation (iV).*

31. Hager, G. and Belhumeur, P. (1996).Real-time tracking of image regions with changes in geometry and illumination. *In Proceedings of IEEE Conference on Computer Vision and Pattern Recognition.,* 403–410.

32. Hai-Xia, X. Yao-Nan, W. Wei, Z. Jiang, Z. and Xiao-Fang, Y. (2011). Multi-object visual tracking based on reversible jump Markov chain Monte Carlo. *IET Computer Vision.,*5(5) : 282–290.

33. Heng, W,J. Ngan, K,N. (1999). The implementation of object-based shot boundary detection using edge tracing and tracking. *Proceedings of the 1999 IEEE International Symposium on Circuits and Systems.,*4(1) : 439-442.

34. Hidetomo, S. (2013). Video based Tracking, Learning, And Recognition Method for Multiple Moving Objects. *IEEE Transactions on Circuits and Systems for Video Technology.,* 23(10) : 1661 – 1674.

35. Hofmann, M. Haag, M. Rigoll, G. (2013). Unified hierarchical multi-object tracking using global data association. *In: IEEE Conference on Computer Vision Pattern Recognition.*, 3682-3689.

36. Hu,W. Li, X. Zhang, X. Shi, X. Maybank, S. and Zhang, Z. (2011) Incremental tensor subspace learning and its applications to foreground segmentation and tracking. *International Journal of Computer Vision.*, 91(3) : 303–27.

37. Huazhong, N. Liang, W. Weiming, H. and Tieniu, T. (2002). Model-based Tracking of Human Walking in Monocular Image Sequences. In *Proceeding of the 2002 IEEE Region 10 Conference on Computers, Communications, Control and Power Engineering.*, 1(1) : 537 – 540.

38. Hui, W. Ting-Zhu, H. Zongben, X. and Yilun, W. (2014). An active contour model and its algorithms with local and global Gaussian distribution fitting energies.*InformationScience.s.*, 263(1) : 43–59.

39. Hui, Z. Baojun, Z. Jianke, L. and Jichao, J. (2010). Visual Tracking for Arbitrary Shaped Object. *2nd International Conference on Future Computer and Communication,* 2(1) : 53-57.

40. Hyunguk, C. and Moongu, J. (2016).Data association for Non-overlapping Multi-camera Multi-object tracking based on Similarity function. In *proceedings of IEEE International Conference on Consumer Electronics.*

41. Jerome, B. Francois, F. Engin, T. and Pascal, F. (2011). Multiple Object Tracking Using K-Shortest Paths Optimization. *IEEE Transactions on Pattern Analysis and Machine Intelligence.*, 33(9) : 1806-1819.

42. Jihao, Y. Chongyang, F. and Jiankun, H. (2012). Using incremental subspace and contour template for object tracking. *Journal of Network and Computer Applications.*, 35(1) : 1740–1748.

43. Junda, Z. Yuanwei, L. and Yuan, F, Z. (2010). Object Tracking in Structured Environments for Video Surveillance Applications. *IEEE Transactions on Circuits and Systems for Video Technology.*, 20(2) : 223-235.

44. Junqiu W. and Yasushi Y. (2008). Integrating Color and Shape-Texture Features for Adaptive Real-Time Object Tracking. *IEEE Transactions on Image Processing.*, 17(2) : 235-240.

45. Kanungo, T. Qigong, Z. (2004). Estimating degradation model parameters using neighbourhood pattern distributions: an optimization approach. *IEEE Transactions on Pattern Analysis and Machine Intelligence.*, 26(4) : 520-524.

46. Kanungo, Tapas, Mount. D,M. Netanyahu, N,S. (2002). An efficient k-means clustering algorithm: analysis and implementation. *IEEE Transactions on Pattern Analysis and Machine Intelligence.*, 24(7) : 881-892.

47. Kass, M. Witkin, A. and Terzopoulos, D. (1988). Snakes: Active contour models. *International Journal of Computer Vision.*, 1(4) : 321–331.

48. Khan, Z. Balch, T,R. Dellaert, F. (2005a). MCMC-based particle filtering for tracking a variable number of interacting targets. *IEEE Transaction on Pattern Analysis and Machine Intelligence.*, 27(11) : 1805–1918.

49. Khan, Z. Balch, T,R. Dellaert, F. (2005b). Multitarget tracking with split and merged measurements. *In Proceedings of IEEE Conference on Computer Vision and Pattern Recognition (CVPR).*,605–610.

50. Kim, T. Lee, S. and Paik, J. (2011). Combined shape and feature-based video analysis and its application to non-rigid object tracking. *IET Image Processing.*, 5(1) : 87 – 100.

51. Kosmopoulos, D, I. Voulodimos, A S. and Doulamis, A, D. (2013). A system for multicamera task recognition and summarization for structured environments. *IEEE Transaction on Industrial Informatics.*,9(1) : 161–171.

52. Kra¨ußling, A. (2008). Tracking multiple objects using the Viterbi algorithm. *Lecture Notes Electrical Engineering.*, 15(1) : 233–247.

53. Kratz, L. Nishino, K. (2012). Tracking pedestrians using local spatio-temporal motion patterns in extremely crowded scenes. *IEEE Trans Pattern Anal Mach Intell.*, 34(5) : 987-1002.

54. Kreucher, C. Morelande, M. Kastella, K. and Hero A, O. (2005). Particle filtering for multitarget detection and tracking. *IEEE Transaction on Aerospace conference.*,1396–1414.

55. Kwak, K. Kim, J,S, Min J. and Park, Y,W. (2014). Unknown multiple object tracking using 2D lidar and video camera. *IET Electronics letter.s,* 50(8) : 600-602.

56. Leibe, B. Schindler, K. Cornelis, N. and Van, Gool, L. (2008). Coupled Object Detection and Tracking from Static Cameras and Moving Vehicles. *IEEE Transactions on Pattern Analysis and Machine Intelligence.*,30(10) : 1683-1698.

57. Li, B. and Chellapa, R. (2000). Simultaneous tracking and verification via sequential posterior estimation. *In Proceedings of IEEE Conference on Computer Vision and Pattern Recognition.*, 110–117.

58. Longyin, W. Zhen, L. Siwei, L. Stan, Z,L. and Ming-Hsuan, Y. (2016) Exploiting Hierarchical Dense Structures on Hypergraphs for Multi-Object Tracking. *IEEE Transactions on Pattern Analysis and Machine Intelligence.*, 38(10) : 1983 – 1996.

59. Luka, C. Matej, K.and Ales, L. (2013). Robust Visual Tracking Using an Adaptive Coupled-Layer Visual Model. *IEEE Transactions On Pattern Analysis And Machine Intelligence.*, 35(4) : 941-957.

60. Maha, M, A. Howida, A, S. and Ashraf, S, H. (2014) .New technique for online object tracking-by-detection in video. *IET Image Processing.*, 8(12) : 794 - 803.

61. Ming, Y. Junsong, Y. and Ying, W. (2007). Spatial selection for attentional visual tracking. *IEEE Conference on Computer Vision and Pattern Recognition.,* 1 – 8.

62. Mohamed, A, N. Omair, Ahmad, M. Swamy, M, N, S. Jongwoo, L. and Ming-Hsuan, Y. (2017). Online multi-object tracking via robust collaborative model and sample selection. *Computer Vision and Image Understanding.,* 154(1) : 94–107.

63. Mohamed, E. Nasreddine, T. Kidiyo, K. and Joseph, R. (2016). Parallel algorithm implementation for multi-object tracking and surveillance. *IET Computer Vision.,* 10(3) : 202 - 211.

64. Mughadam B., and Pentland A. (1997) .Probabilistic visual learning for object representation. *IEEE Transaction on Pattern Analysis and Machine Intelligence.,* 19(7) : 696–710.

65. Murray, D. and Basu, A. (1994). Motion tracking with an active camera. *IEEE Transaction on Pattern Analysis Machine Intelligence.,* 16(5) : 449–459.

66. Okuma, K. Taleghani, A. De, Freitas, N. (2004) .A boosted particle filter: multitarget detection and tracking. *In Proceedings of IEEE European Conference on Computer Vision (ECCV).,* 28–39.

67. Osher, S. and Sethian, J, A. (1988). Fronts propagation with curvature dependent speed: Algorithms based on Hamilton-Jacobi formulations. *Journal of Computational Physics.,* 79(1) : 12–49.

68. Paragios, N. and Deriche, R. (2000). Geodesic active contours and level sets for the detection and tracking of moving objects. *IEEE Transaction on Pattern Analysis and Machine Intelligence., 22(3) : 266–280.*

69. Paragios, N. and Rousson, M. (2002) .Shape priors for level set representations.*ComputerVision.,*78–92.

70. Pätzold, M. Sikora, T. (2011). Real-time person counting by propagating networks flows.in: *AVSS.,*66–70.

71. Peihua, L. Tianwen, Z. (2003).Visual Contour Tracking Based on Particle Filters.*Image and Vision Computing.,* 21(1) : 111-123.

72. Portela, Sotelo, M, A. Desseree, E. Moreau, J,M. Shariat, B. and Beuve M.(2012) 3-D Model-Based Multiple-Object Video Tracking for Treatment Room Supervision. *IEEE Transactions on biomedical engineering., 59(2) : 562-570.*

73. Richard J.D. Moore, Gavin J. T., Angelique C. P., Thomas P., Bruno van S. and Mandyam V. S. (2014). FicTrac: A visual method for tracking spherical motion and generating fictive animal paths. *Journal of Neuroscience Methods, Vol.225, 106–119.*

74. Rui, Yao. (2015). Robust Model-Free Multi-Object Tracking with Online Kernelized Structural Learning. I*EEE transaction on signal processing letters.,*22(12) : 2401-2405.

75. Samarjit, D. Amit, K. and Namrata, V. (2012). Particle Filter With a Mode Tracker for Visual Tracking Across Illumination Changes.*IEEE Transactions On Image Processing.,*21(4) : 2340-2347.

76. Sanghoon, K. Sangmu, L. and Seungjong, K. (2008).Object Tracking of Mobile Robot using Moving Color and Shape Information for the aged walking. *Proceedings of the Second International Conference on Future Generation Communication and Networking.,* 02(1) : 293-297.

77. Satrughan, K. and Jigyendra, Sen, Y. (2016). Video object extraction and its tracking using background subtraction in complex environments. *Perspectives in Science.,*8(1) : 317–322.

78. Sayed, Hossein, K. and Ivan, V, B. (2013).Video Object Tracking in the Compressed Domain Using Spatio-Temporal Markov Random Fields. *IEEE Transactions on Image Processing.*, 22(1) : 300-313.

79. Serby, D. Koller-Meier, S. and Gool, L,V. (2004). Probabilistic object tracking using multiple features. *In proceedings of IEEE International Conference of Pattern Recognition (ICPR).,* 184–187.

80. Serhan, G. Jan, Timo, M. Cornelius, He. Thomas, S. and Wojciech, S. (2016).Hybrid Video Object Tracking in H.265/Hevc Video Streams. In *proceedings of IEEE 18th International Workshop on Multimedia Signal Processing (MMSP).*

81. Shao-Yi, C. Wei-Kai, C. Yu-Hsiang, T. and Hong-Yuh, C. (2013). Video Object Segmentation and Tracking Framework With Improved Threshold Decision and Diffusion Distance. *IEEE Transactions on Circuits and Systems for Video Technology.,* 23(6) : 921-934.

82. Shinsuke, Y. Hirotsugu, O. Kazuo, I. and Tetsuya, Y. (2016). Real-time object tracking based on scale-invariant features employing bio-inspired hardware.*NeuralNetworks.,* 81(1) : 29–38.

83. Shu, T. Fei, Y. Gui-Song, X. (2016). Multi-object tracking with inter-feedback between detection and tracking. *Neurocomputin.g, 171(1) : 768–780.*

84. Shucheng, H. Shuai, J. and Xia, Z. (2016). Multi-object tracking via discriminative appearance modelling. *Computer Vision and Image Understanding.,* 153(10) : 77–87.

85. Song, X. Cui, J. Zha, H. and Zhao, H. (2008). Vision-Based Multiple Interacting Targets Tracking via On-Line Supervised Learning. In *Proceedings of 10th European Conference on Computer Vision.,* 3(1) : 642-655.

86. Soo-Chang, P. Ching-Min, C. (1999). Color image processing by using binary quaternion-moment-preserving thresholding technique. *IEEE Transactions on image Processing.,*8(5) : 614-628.

87. Swati, N, S. Ajitkumar, K. and Dilip, M. (2016).Multi-Object Tracking Using TLD Framework. In *proceedings of IEEE International Conference on Recent Trends in Electronics Information Communication Technology.*

88. Tao, H. Sawhney, H. and Kumar, R. (2000). Dynamic layer representation with applications to tracking. In *Proceedings of IEEE Conference on Computer Vision and Pattern Recognition.,* 2(1) : 134–141.

89. Tao, W. Qian, D. Yimin, Z. Gang, S.Chunrong, L. and Gary, B. (2004).A Dynamic Bayesian Network Approach to Multi-cue based Visual Tracking. *Proceedings of the 17th International Conference on Pattern Recognition.,*2(1) : 167-170.

90. UCSD datasets http://www.svcl.ucsd.edu/projects/anomaly/dataset.htm.

91. Veenman, C.Reinders, M. and Backer, E. (2001). Resolving motion correspondence for densely moving points. *IEEE Transaction on Pattern Analysis and Machine Intelligence.,* 23(1) : 54–72.

92. Vermaak, J. Doucet, A. and Perez, P. (2003).Maintaining Multimodality through Mixture Tracking. *Proc. IEEE Int'l Conf. Computer Vision., 1110-1116.*

93. Wang, H. and Suter, D. (2005). Tracking and Segmenting People with Occlusions by a Sample Consensus-Based Method. *In Proceedings of IEEE International Conference on Image Processing.,* 2(1):410-413.

94. Wang, Y. Wu, J.Huang, W. and Kassim, A. (2007) .Gaussian mixture probability hypothesis density for visual people tracking. In *Proceedings of 10th International Conference on Information Fusion.,* 1–6.

95. Weiming, H. Xi, L. Wenhan, L. Xiaoqin, Z. Stephen, M. and Zhongfei, Z. (2012). Single and Multiple Object Tracking Using Log-Euclidean Riemannian Subspace and Block-Division Appearance Model. *IEEE transactions on pattern analysis and machine intelligence.,* 34(12) : 2420-2440.

96. Weiming, H.Xue, Z. Min, H. and Steve, M. (2009). Occlusion Reasoning for Tracking Multiple People. *IEEE Transactions on Circuits and Systems for Video Technolog.y,* 19(1) : 114-121.

97. Weiming, H. Xue, Z. Wei, L. Wenhan,L. Xiaoqin, Z. and Stephen, M. (2013). Active Contour-Based Visual Tracking by Integrating Colors, Shapes, and Motions. *IEEE Transactions On Image Processing.,*22(5) : 1778-1792.

98. Wenhan, L. Junliang, X. Xiaoqin, Z. Xiaowei, Z. Tae-Kyun K. (2015). Multiple Object Tracking: A Literature Review, Computer Vision and Pattern Recognition.

99. Wu-Chih, H. Chao-Ho, C. Tsong-Yi, C. Deng-Yuan, H. and Zong-Che W. (2015) .Moving Object Detection and Tracking from Video Captured by Moving Camera. *Journal of Visual Communication and Image Representation.,*30(12) 164–180.

100. Xiao, L. Dacheng, T. Mingli, S. Luming, Z. Jiajun, B. and Chun, C. (2014). Learning to Track Multiple Targets. *IEEE transactions on neural networks and learning systems.,*26(5) : 1060 - 1073, 2014.

101. Xiaokai M., Jin C., Tianli H., and Zhongyi W. (2016) .A Video Object Tracking Algorithm Combined Kalman Filter And Adaptive Least Squares Under Occlusion. In *proceedings of International Conference on Image and Signal Processing, Bio Medical Engineering and Informatics (CISP-BMEI).*

102. Xiaolong, Z.Youfu, L.Bingwei, H. and Tianxiang, B. (2014) .GM-PHD-Based Multi-Target Visual Tracking Using Entropy Distribution

and Game Theory. *IEEE transactions on industrial informatics., 10(2) : 1064-1076.*

103. Yilmaz, A. Li, X. and Shah, M. (2004).Contour based object tracking with occlusion handling in video acquired using mobile cameras. *IEEE Transaction on Pattern Analysis and Machine Intelligence., 26(11) : 1531–1536.*

104. Ying, W. and Thomas, S, H. (2004). Robust Visual Tracking by Integrating Multiple Cues based on Co-inference Learning. *International Journal of Computer vision.,*58(1) : 55-71.

105. Yonatan, T, T. Eyasu, Z. Marcello, P. and Andrea, P. (2016). Multi-object tracking using dominant sets. *IET Computer Vision., 10(4) :*289 – 297.

106. Zhen, D. and David, V. (2006).A modified Murty algorithm for multiple hypothesis tracking. *Signal and Data Processing of Small Targets.*

107. Zhenhua, G. Zhang, L. Zhang, D. (2010). A Completed Modeling of Local Binary Pattern Operator for Texture Classification. *IEEE Transactions on image processing., 19(6) : 1657-1663.*

108. Zhenyu, H. Xin, L. Xinge, Y. Dacheng, T. and Yuan, Y, T. (2016) .Connected Component Model for Multi-Object Tracking. *IEEE Transactions on Image Processing., 25(8) : 3698 - 3711.*

109. Zhong, W. Altun, G. Harrison, R. Tai, P,C, and Yi, Pan. (2005) .Improved K-means clustering algorithm for exploring local protein sequence motifs representing common structural property. *IEEE Transactions on NanoBioscience.,*4(3) : 255-265.

110. Zhou, X. Hu, W. Chen, Y. and Hu, W. (2007).'Markov random field modelled level sets method for object tracking with moving cameras. In *Proceedings of Asian Conference on Computer Vision.,*832–842.

Printed by Books on Demand GmbH, Norderstedt / Germany